The Weather Machine

Nigel Calder

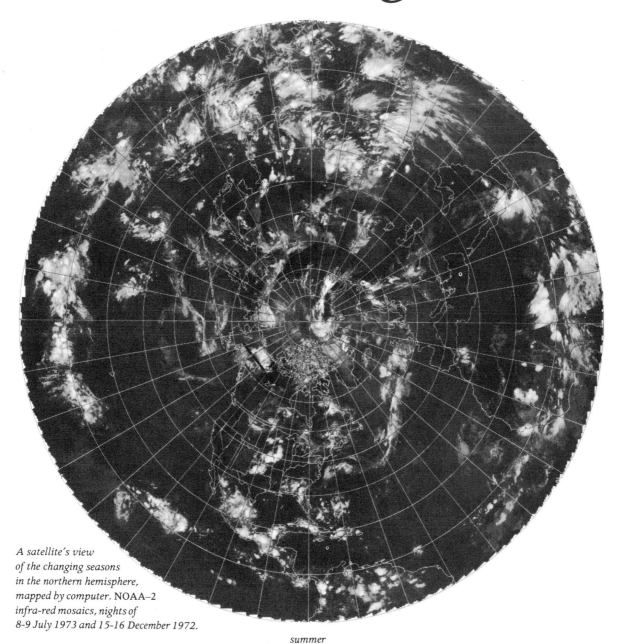

*A satellite's view
of the changing seasons
in the northern hemisphere,
mapped by computer. NOAA–2
infra-red mosaics, nights of
8-9 July 1973 and 15-16 December 1972.*

summer

The Weather Machine

THE VIKING PRESS   NEW YORK

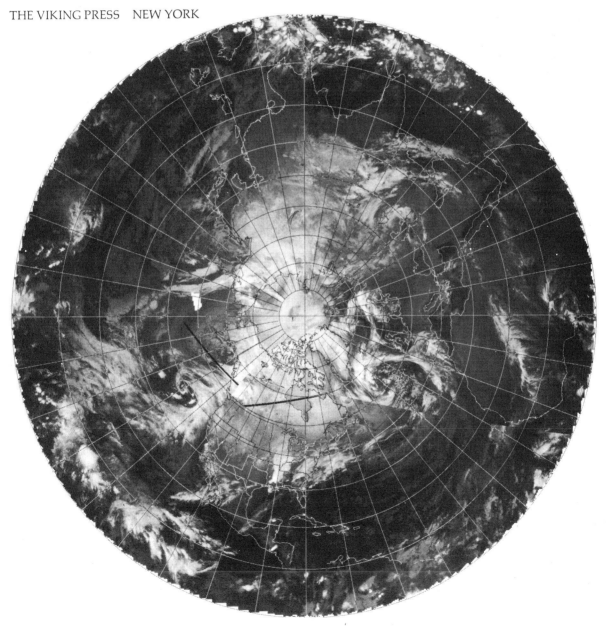

winter

Author's note

I have been able to travel around the world, and to meet leading experts on weather and climate in a dozen countries. For that opportunity I am grateful to the BBC and its overseas coproducers for the television programme *The Weather Machine*. This book draws and enlarges upon information gathered for the programme, but I have conceived and written the book quite separately from the television script. My gratitude is even greater to the experts whom we consulted and who never hesitated to give us information and advice. As there are nearly two hundred of them, I cannot acknowledge them all by name. Still less can I deal adequately with all their diverse and fascinating lines of research in the pages that follow, but they have all contributed to the general perspective of the book. Any shortcomings are strictly my responsibility.

Copyright © 1974 by Nigel Calder
All rights reserved
Viking Compass Edition
issued in 1976 by The Viking Press, Inc.
625 Madison Avenue, New York, N.Y. 10022

LIBRARY OF CONGRESS CATALOGING IN PUBLICATION DATA
  Calder, Nigel.
    The weather machine.
    "Viking compass edition."
    Bibliography: p.
    Includes index.
    1. Climatic changes. I. Title.
  QC981.8.C5C34 1976   551.6'365   75-35790
  ISBN 0-670-00618-1 -ppbd.
        670-75425-0 -hdbd.

Printed in U.S.A.
Color sheets printed in England

Wind speeds are given in knots (nautical miles per hour)

knots:	10	20	50	100	579
miles per hour:	11·5	23	57·6	115	689

Temperatures are given in degrees C.

°C:	−80	−40	0	10	20	30	40	100
°F:	−112	−40	32	50	68	86	104	212
			(ice melts)					(water boils)

A *difference* of $1°C = 1·8°F$.

Contents

The 120-minute television programme, *The Weather Machine*, was first transmitted on BBC2 on 20 November 1974. It was made by the BBC as a coproduction with WNET (New York), Sveriges Radio (Stockholm), KRO (Hilversum), OECA (Toronto) and ZDF (Mainz). The executive producer was Alec Nisbett, assisted by Robin Bootle, who was also the studio director. John Baker was the principal film cameraman and the film editor was Christopher Woolley. Graphics were by Charles McGhie and visual effects by Michealjohn Harris. The studio designer was Stewart Marshall. The programme was presented by Magnus Magnusson, narrated by Eric Porter, and written by Nigel Calder.

Looking for water. Ethiopian villagers afflicted by the recent drought dig into the dry bed of a river.

1 Change in the Wind

When the former British prime minister spoke with uncustomary erudition about the weather of 1972-3, his opponents thought he was clutching at any lame excuse for the high price of food. He told of a lack of snow on the steppes that brought about a poor harvest in the Soviet Union and China. Parliament was beside itself with laughter. As Ted Heath went on to mention the frost in Brazil which hurt the coffee crop, the hilarity redoubled. Several lawmakers were too overcome by mirth to catch his remark about fishmeal supplies for animal feeding stuffs that had suffered from a very sudden change in direction of the current off Peru. The prime minister had misjudged the manner of his homily. Yet the events he related were indeed typical of the meteorological upsets of the early 1970s. The world's larder of grain reserves became almost bare. In 1974 floods in the wheatfields of Canada, spring droughts across much of northern Europe and in the Soviet grainlands, and summer drought in the US Midwest threatened the loss of yet more harvests.

Many people without these strategic worries, but concerned for their business or leisure interests, have sensed for quite a while that something unfamiliar is happening to the weather. At one time, remember, the custom was to blame its quirks on the testing of H-bombs. Only ten years ago I had difficulty in finding any experts except for a few oddballs who would allow that the weather was getting worse. The 'oddballs' were pioneering researchers into the history of the climate. They were the readier to admit changes in our own time because they knew the weather never had stayed quite the same, from one decade to the next.

And the world was utterly different when mankind was young and lived by its arrows and its wits. For practical purposes it was bounded to the north by the frigid deserts of England, Holland and central Germany. Beyond lay sheets of solid ice, a mile thick. Where Chicago, New York, Hamburg and Moscow stand today, the buried land sagged under its white burden. The sea was low and hunters could and did cross dryshod from Siberia to Alaska. Wandering through the Americas, they had to skirt the glaciers of California and Mexico. The great thaw came only ten thousand years ago. Humanity prospered and spread its civilisation even into the Arctic. Now, very recent discoveries imply that the chance of the next ice age starting in our own lifetimes is not zero.

This report on current research into the changing weather is no work of science fiction, despite its currency of strange disasters and its men and women who confront the elements with computers and satellites, microscopes and fireworks. Nor is it yet another chronicle of mankind's ecological misdeeds. Foolish we may be, but there are mysterious natural forces at work, probably surpassing our own strenuous efforts to blot out the Sun by pollution. Least of all do I set out to scare anyone or to 'startle the bourgeoisie'. For one thing, the non-stop movie of the weather would be a fascinating enough subject even if it did not rule our livelihood and habits; even if the show were not changing. The variations in climate, too, have intrinsic interest because they helped to shape our landscapes, our evolution and our history. Progress in meteorology and climatology, and the remarkable international cooperation that underpins it, give solid grounds for optimism even though some of the discoveries seem dire. But the chief reason for not setting out to prove catastrophe is precisely that the situation is already grave.

Many people, in Africa for example, are already dying as a result of a moderate change of climate in our time. The effects of a major climatic change would be comparable with a nuclear war. Decisions of life and death have to be taken on the basis of imperfect knowledge and in the face of conflicting predictions about

the climate of the next few decades. By discouraging action that could avert human suffering and loss of life, unnecessarily pessimistic forecasts may be as harmful as unduly optimistic ones. So my emphasis is on reporting the facts (many of them new or only newly understood) and the ongoing efforts to explain why the weather machine can change gear. As reasons for optimism or pessimism come to hand, I shall compare them.

Snow and ice show up brightly on satellite pictures of the Earth. You can be rid of the clouds by taking the least brightness at each point over a period of days, so that only permanent brightness of the snow and ice remains. In 1967, the US National Oceanic and Atmospheric Administration began issuing weekly maps of snow and pack-ice over the Earth, which varies of course from season to season. In 1971-3 the snow and pack-ice cover in the northern hemisphere formed earlier in the winter and covered an area eleven per cent larger than it did in 1970. The meteorological records show a progressive cooling in the northern hemisphere since around 1950, especially in the Arctic. Although the meaning of these figures is queried by some experts, the Russians are building icebreakers as fast as they can, to keep open their northern shipping lanes.

More subtle differences between the present period and the first half of this century, all over the world, tell of general weakening of the winds on which many places depend for rain and warmth. One symptom is a marked reduction in the frequency of the 'prevailing' south-west winds of Britain. Simple arguments about links between one part of the weather machine and another quite plausibly relate the African drought to a cooling in the north. These are a few of the reasons why there are fears that the Earth's climate may be changing substantially and for the worse.

The weather is a system intricate in both space and time. Whether sunshine or rain falls on you today de-

pends on what was happening in the sea off Japan last week, in the Indian Ocean last month, in the sea ice off Iceland last winter, in the polar ice sheets a century ago, in the Earth's orbit around the Sun ten thousand years back, in the movements of continents ten million years ago. Any route through the system is an arbitrary one. This first chapter puts the climatic oddities of our own time into the context of changes over the past centuries. Chapter 2 will be concerned with possible causes for the changes and with the present efforts of meteorologists to understand the intricacies of the weather machine. The third chapter chronicles the astonishing new discoveries about the ice ages and evaluates the threat of a major cooling in our time.

Climates in deep-freeze

In the summer of 1974, an international party was at work on the high ice sheet of Greenland. It was at a map-reference named Crête where the ice is ten thousand feet thick. Researchers from Copenhagen and Berne were in attendance while American army engineers were trying out new techniques for drilling into the ice. They were extracting, piece by piece, a continuous sample – a 'core' – of ice, down to a considerable depth below the surface. Their main aim was to recover a core 1200 feet long. That would give the Danes about 1400 years of history to work on. The Danes were expert in telling the climate of past centuries from a piece of ice. The Swiss were at Crête to try to improve the methods of dating. They would melt the ice deep down to release the air from the little bubbles that make ice white; back in Berne, a measurement of the radioactive carbon dioxide in the air would show how long ago the ice was formed.

The snows that fall each year on the ice sheets of Greenland and Antarctica are preserved in natural

deep-freeze. As new snow falls its weight crushes the underlying snow into ice, but it does not upset the orderly layering of successive years' contributions to the ice sheet – the deeper, the older. If the drilling site is carefully chosen, the layers should have remained scarcely disturbed even by movements of the mass of ice. The Crête expedition was continuing research that began in 1966 when the US Army succeeded in drilling through 4500 feet of ice, right to the underlying rock, at a site called Camp Century, in northern Greenland. It provided a record going back an estimated 150,000 years, to before the start of the most recent ice age. A similar, even deeper core, to the base of the West Antarctic ice sheet at Byrd Station, was forthcoming in 1968. But Danish investigators of the long Camp Century core, in particular, obtained results so promising for the study of changes of climate that it became important to drill more holes at different locations to check the reliability of their deductions. The Crête operation was just one of several short-core operations called for in the Greenland Ice Sheet Program, before the next attempt to drill to the bedrock. The Russians have begun similar operations at Vostok, deep in the interior of the great ice sheet of East Antarctica.

In short, Willi Dansgaard and his colleagues at the University of Copenhagen have started a fashion. Dansgaard is a physicist who became interested in using machinery for weighing atoms as a means of tracing the climatic conditions in which snow fell, hundreds or thousands of years ago. Half-sections of the 1966 Camp Century core went to Copenhagen, where Dansgaard was able to show the potency of the technique, as he traced the great variations in climate through the most recent ice age and the lesser but very important changes of the past 10,000 years. Today the work continues feverishly, while the Americans supply Dansgaard with new ice cores as fast as they can.

The idea is to measure, in small pieces of the core, the occurrence of the heavy kind of oxygen atoms – oxygen-18. The snow was made of water that evaporated from the Atlantic and travelled to Greenland. Molecules of sea water containing the commoner form of oxygen, oxygen-16, are lighter and evaporate more readily, and remain evaporated longer than do the small minority that contain heavy oxygen. In colder air the heavier molecules have even less chance of remaining as vapour; they tend to fall out as rain or snow before they reach Greenland. The proportion of heavy oxygen in the ice therefore gives a general indication of the climate prevailing when the snow fell on Greenland – the less heavy oxygen there is, the colder the climate was. A machine called a 'mass spectrometer', in which a magnet sorts atoms according to their weights, can measure the heavy oxygen very accurately.

In Dansgaard's laboratory in Copenhagen a cold room preserves the Greenland cores. Small pieces are cut out and melted for analysis. Each sample of melted water spends a night being shaken with carbon dioxide gas in a small bottle, so that it can exchange atoms with the gas. By next morning the gas possesses a good representation of the oxygen atoms from the Greenland ice. Then it is the work of a minute or so to run the carbon dioxide through the mass spectrometer and find out the proportion of heavy oxygen. With a growing supply of ice cores awaiting analysis, Dansgaard has made the work automatic; under computer control, the machinery copes with 256 pieces of ice from a core every day.

The long core from Camp Century clearly showed the deep, long-lasting dip in the proportion of heavy oxygen in the ice through the period of the most recent ice age – 70,000 to 10,000 years ago. Dansgaard and his colleagues also found a lot of detail in the record showing fluctuations between very cold and less cold condi-

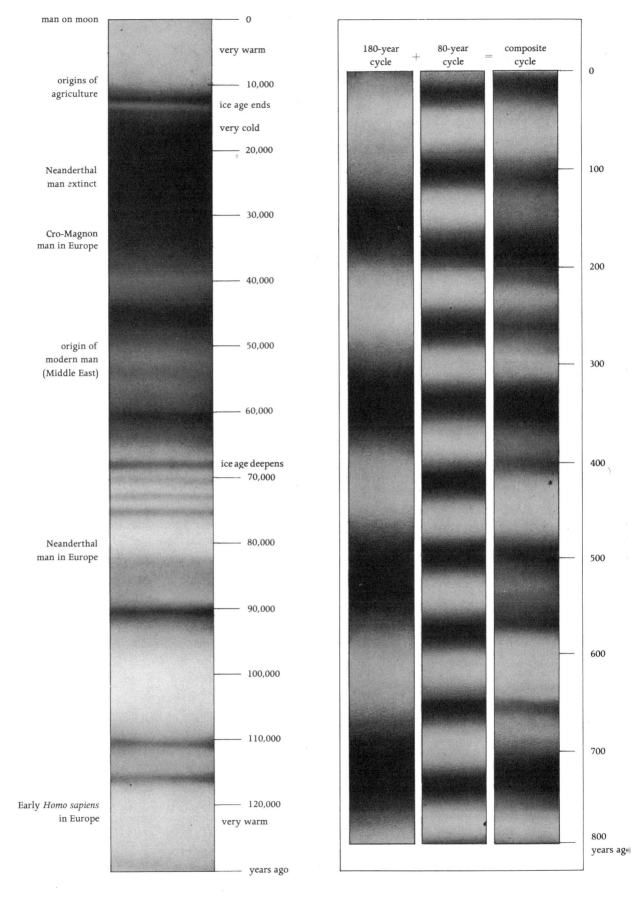

man on moon — 0

very warm

origins of
agriculture — 10,000

ice age ends

very cold

— 20,000

Neanderthal
man extinct

— 30,000

Cro-Magnon
man in Europe

— 40,000

origin of
modern man
(Middle East) — 50,000

— 60,000

ice age deepens
— 70,000

Neanderthal
man in Europe — 80,000

— 90,000

— 100,000

— 110,000

Early *Homo sapiens*
in Europe — 120,000

very warm

years ago

180-year cycle + 80-year cycle = composite cycle

0

100

200

300

400

500

600

700

800
years ago

tions during the ice age itself, and between warm and cooler conditions in the generally warm periods before and since the ice age. Notable peaks of warmth occurred, for example, around AD 1100 and 1930. These fluctuations correspond with exceptionally good or bad conditions experienced and recorded by our forefathers. Since 1930, the proportion of heavy oxygen has fallen sharply.

By looking closely at the icy record of the past eight centuries the Danish group found two apparent rhythms of change interwoven with one another – cycles of about 80 years and 180 years which they ascribed to fluctuations in the Sun. The exceptional warmth of the early 20th century was due, on this interpretation, to the warm phases of these two cycles coinciding, and reinforcing one another. Running the cycles into the future led to predictions. When the Copenhagen scientists offered their prediction in 1970, they contented themselves with the next 50 years and were careful to point out that pollution and other effects of human activities could alter the course of the climate. Nevertheless, the natural trend – if their cycles were real – was marked cooling continuing into the 1980s, reversing but not by very much until about 2015, and then cooling again. Even at the peak of the reversal the climate would be much cooler than in the 1930s. On a longer time-scale the cycles threatened several centuries of cooler conditions, similar to those that prevailed in recent centuries – in the period called the Little Ice Age.

Although Dansgaard has been anxious to say that the prediction is not to be taken too seriously, the need to assess its possible reliability is a powerful incentive in the continuing work with the Greenland ice. The chief difficulty has been in dating the layers of ice. At first Dansgaard based his dates simply on the depth of the ice, assuming steady snowfall and making allowance

for the squeezing and spreading of the lower layers under the great weight of overlying ice. But there are discrepancies from core to core. One method the Danes are using to check the dating, in the younger ice, is by finer analysis to pick out the seasonal variations within each year. They can then count the years as one counts the rings of a tree. They are also looking to volcanic eruptions that would have shed some of their dust; one of Dansgaard's young colleagues is testing melted samples of the Greenland ice with an instrument capable of counting the small dust particles they contain. Swiss work, on dating the ice by the radioactive carbon in its bubbles, has already been mentioned. The pressure is on, to see whether the rhythms are real and the climatic outlook is as bleak as it seems.

An earlier but similar forecast also told of a worsened climate coming in the north, to last a hundred years or more. The forecast emanated from Finland in 1963 and was based on quite different information, the rate of growth of trees at the chilly forest limits of Lapland, as judged by the widths of the yearly rings in the trunks of living and dead trees. This natural record went back to the 12th century AD. It showed periods of warmth or cold, when the trees grew well or poorly. The mathematical analysis suggested rhythms of 72, 92 and 204 years, not so very different from the 80-year and 180-year cycles in the heavy oxygen of the Greenland ice. Another prognosis, over a longer time-scale, comes from other icy records.

Glaciers, the slow-moving rivers of ice that form beside snow-capped mountains, are climatic thermometers for their regions. In warm periods they retract their snouts up the valleys. In cold periods, either because there is more snow in winter or less melting, owing to cooler weather in summer, the glaciers advance. During much of the 20th century many northern glaciers have been melting away, but the retreat has slowed

The Rhône glacier in Switzerland (left) as a thermometer of climate, showing its retraction during the warm decades earlier in this century. Below: Crête, in Greenland, 1974. Drilling operations recovered new samples of the ice sheet, for pursuit of climatic history.

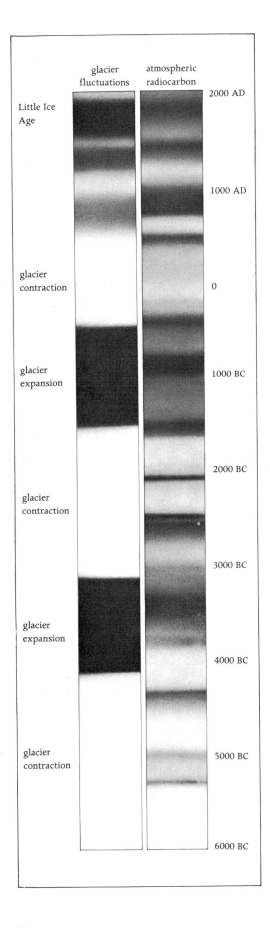

glacier fluctuations	atmospheric radiocarbon	
Little Ice Age		2000 AD
		1000 AD
glacier contraction		0
glacier expansion		1000 BC
		2000 BC
glacier contraction		3000 BC
glacier expansion		4000 BC
glacier contraction		5000 BC
		6000 BC

The comings and goings of the glaciers are shown in the left-hand column. Black denotes the advance of the glaciers. An apparent rhythm of 2500 years implies that we are still in a generally cold phase. A similar pattern is discernible in the natural production of radioactive carbon in the Earth's atmosphere (nearer column), suggesting that the Sun changes in the same rhythm – see page 76. (After G. Denton and W. Karlén.)

down recently. The Antarctic glaciers, unlike their northern counterparts, have not changed significantly in the last fifty years or so. The evidence comes from lichen growing on rocks and boulders around the glaciers. Lichen can live right down to the ice surface but it is incredibly slow to grow – a patch of lichen takes 500 years to become four inches wide. Below the glaciers of Swedish Lapland a 'tidemark' of lichen reveals the last icy maximum (1916) with new lichen only slowly establishing itself on the rock now laid bare. Beside Antarctic glaciers, on the other hand, there is no gap between the lichen and the ice, showing that there has been no retreat in that region.

Swedish researchers use lichen for dating the changes in their northern glaciers. But the general technique for deducing an advance of the ice in the past is to identify an old 'moraine'. That is a heap of stones dumped at the former position of the snout of a glacier and marking its greatest extent at that time. Then you find logs or other plant matter, in the moraine or nearby, which can be dated by the amount of radioactive carbon in them.

George Denton of the University of Maine is one of the new generation of ice geologists who have been very busy for a decade studying the histories of glaciers in Swedish Lapland, in Alaska and the adjacent part of Canada, and in Antarctica. He sees in the northern glaciers a dominant rhythm that quite overrides the lesser fluctuations of a century or two, of the kind reported by Dansgaard. The glaciers tell a fairly consistent story of successive advances and retreats – of cold and warm periods – since the big melt at the end of the most recent ice age. The most important advances in the glaciers have reached a maximum every 2500 years, with long warm periods of glacier retreat in between.

Here I have to admit an early setback for my policy of giving grounds for optimism wherever possible. Denton's initial intimations were most promising. You could say that having emerged, just in this century, from one of Denton's cold periods we might look forward to a thousand years of improving conditions. I am sorry to say that Denton has recently come to the conclusion that we are still in the cold period, which may have a few more centuries to run.

With no conflict between Denton's big cycle and Dansgaard's little ones we can escape their implications for the natural trend only by declaring the cycles to be illusory. That is possible but, as we shall see, the rhythms have an apparent physical origin in the changing climate of the Sun itself. Less elusive than the cycles, and the chilly future they may or may not imply, are the general facts of past climatic change.

Little Ice Age

The ice floating in the sea off Greenland and Iceland is at present impeding shipping and fishing more than it has done in living memory – but less, so far, than it did in the 17th century, in the depths of the Little Ice Age. Then the inhabitants were sometimes completely besieged by the ice. Throughout its existence our species has contended with repeated changes of climate. Communities would adapt their hunting habits, their farming practice or their cities to the prevailing climate, only to find after a century or two a change bringing hardship or even forcing them to abandon their lands or die.

The most recent ice age began to expire 12,000 years ago, but deceitfully. Trees and animals moved north, and up the glacier valleys, in the wake of the retreating ice as the world grew steadily warmer for about 1000 years. Then the ice had its last big fling. Quite suddenly the glaciers stopped melting away and advanced again, annihilating the impudent forests. Only after that episode did the world begin to settle down, about 10,000 years ago, to something close to present conditions.

The agriculture of the Middle East and the Indus Valley began just then. In the radically altered pattern of global climates, food supplies from hunting and gathering in those regions may have become less reliable than they were during the ice age. Thereafter the climatic conditions allowed agriculture to spread steadily westwards and northwards across Europe, from its Middle Eastern origins, without any noticeable interruptions, reaching the North Sea around 3000 BC.

As if on the rebound the climate in the first few thousand years after the ice age was very warm – generally warmer than anything experienced since. By 5000 BC Europe was two or three degrees warmer than it has been even during the warmest decades of this century. The floating Arctic ice shrank to a very small area, while the melting of the land ice was raising the sea level. Heavy rainfall in northern Europe swamped many forests and turned them into peat bogs. In that 'climatic optimum' vigorous winds also brought summer rain into what is now the deep Sahara Desert.

Since then the story has been one of overall cooling, a slow decline that is already leading us inexorably into the next ice age. Warm periods of lessening intensity have alternated with cool periods of deepening chill. One of the cool periods (4000-3000 BC) coincided with the rise of the first cities in Mesopotamia and the founding of the first Egyptian dynasty. Glaciers are known to have advanced in Switzerland, Washington State and Patagonia. Life was hard for people in the northern lands. But after 3000 BC, when conditions were altogether warmer again, came the climax of the Megalithic civilisation in north-west Europe, of which Stonehenge in England is a celebrated monument.

The next long cool spell ran from 1300 to 500 BC. Glaciers that had melted completely away in the American Rockies began to reform, and glaciers everywhere advanced: Alaska, Utah, Sweden and Patagonia give definite evidence. The widespread effects of climatic change may well have had a lot to do with the human upheavals of that time, which saw great migrations and invasions. The civilisations of the Hittites and Mycenae fell; those of the Assyrians, the Phoenicians and the Greeks rose. But as always in human affairs there were many other factors, not least the technological transition from the bronze to iron, and the devastation of the Aegean seaports by the enormous volcanic explosion of Thera (Atlantis). Over-zealous attempts to explain history simply by climatic changes may be rash. For one thing it is too easy: you can attribute invasions either to a worsening climate that makes people desperate or to improving conditions that give them the surpluses required for war.

Nevertheless, as a matter of context, the empires of Ashoka in India, of the Chhin dynasty in China, and of the Romans in Europe, North Africa and the Middle East all grew during the warmer period that began around 500 BC. The Romans' rivals in North Africa and the Middle East may have been weakened by generally dryer conditions, and later improvements in the north can only have helped to nurture the 'barbarians' who eventually smashed the Roman Empire. Prosperity and power shifted northwards in Europe, culminating in the Viking age of AD 800-1200. Confusingly, though, for the historian of climate, there was a spell of cooling, AD 700-900, when glaciers advanced, certainly in Alaska. But the latter part of this period was very warm: the glaciers were well retracted and in North America, for example, forests grew much further north than they do today. The American Midwest was warm and dry but, by sharp contrast, China and Japan were cool. Invigorated winds were sweeping them with the cold Siberian air that might otherwise have come to eastern Europe.

The Vikings had erupted like locusts from their newly warmed Scandinavian bases in the 9th century AD, when they invaded Russia, France and Britain. Iceland was settled in 874 and the Viking seamen discovered Greenland in 982; their supposed voyages to North America are still controversial but there was no climatic reason why they should not have easily made the crossing from Greenland. The Mongols mastered much of Asia and poured westwards into Russia and central Europe at the close of the warm period, in the 13th century. The sea ice around Iceland was asserting itself by 1300 and the ensuing century in Europe was a period of cooling but erratic climate. Widespread famines occurred in western Europe in 1315 when heavy summer rainfall reduced the landscape to a sea of mud. The winters remained fairly mild but poor summers persisted. It was a terrible century, with the Black Death halving the population of many districts of Europe and desperate peasants in revolt in France and England. The Alaskan glaciers began creeping forward during that century.

The year in which the English burnt Joan of Arc, 1431, ended in a winter so severe that it can well be said to mark the onset of the Little Ice Age in Europe. As if to demonstrate that the balmy Viking age had ended, every river in Germany froze that winter and many French vines were killed by the frost. During January and especially February 1432, a great 'block' of settled weather over northern Scandinavia swept a current of cold air across Europe from the east. Thereafter cold winters were common and by 1492 the Pope was complaining that no bishop had been able to visit Greenland for 80 years, on account of the ice. He did not know that his Norse congregations had died in the cold by 1450.

Historians of climate differ in their dating of the Little Ice Age, for the very good reasons that a long cold period was punctuated by warmer decades and the effects differed from place to place. I am taking it as the

Hungry and thirsty animals at the Sahara's southern edge. Drought has been the most alarming climatic theme of the early 1970s.

Holland in the Little Ice Age, shown in a detail from an early 17th-century painting by Hendrik Avercamp.

period 1430-1850, which includes almost everyone's opinions. The clearest markers, the glaciers of Europe, advanced decisively in the early 17th century, when villages near Chamonix in France were overwhelmed by ice; again in 1643-53, which was the period of severest winters in western Europe since the end of the most recent ice age; and also in the 1740s, at the time of the war of the Austrian succession and of Bonnie Prince Charlie's Scottish rebellion. Although the Little Ice Age was 'little' in both severity and duration compared with the true ice ages, it brought great suffering in Scandinavia, Scotland, Iceland and New England. Others with full stomachs could enjoy winter sports on frozen rivers and lakes while, during the American Revolution, British troops were able to slide their guns across the ice from Manhattan to Staten Island.

The bald phrase 'Little Ice Age' conceals an enormous variability in both space and time. Abundant historical evidence about the Little Ice Age makes obvious the dangers of vague generalisations about earlier periods – or future periods, for that matter. Winters in China and Japan were actually milder during Europe's Little Ice Age, because weaker winds from the west allowed them to benefit more from the warm Pacific air. The American Midwest was cooler and wetter than before or after the Little Ice Age, because the wet Pacific winds travelled further south across the country. And we can contrast, within this period, extremely wet periods in England, such as the 1560s, with the summer droughts of the 1660s. In 1666, with the River Thames barely navigable for lack of water, plague-ridden London was as dry as kindling and the Great Fire was no surprise.

Least of all in the war-torn and diseased Europe of the Little Ice Age should one be casual about explaining history in terms of weather. But for one notable event, the French Revolution, the link is plain. The worst of the Little Ice Age was past, and in northern France in 1788 May, June and July were excessively hot and the grain shrivelled. On 13 July, just at harvest time, a severe hailstorm added to the losses of many farmers. From that bad harvest of 1788 came the bread riots of 1789, Marie Antoinette's alleged remark 'Let them eat cake', and the storming of the Bastille.

The slow climb out of the worst of the Little Ice Age had begun, but not until the 1860s did the glaciers begin melting conspicuously. Even then there was a further re-advance to come at the beginning of the 20th century, interrupting the general retreat. But after 1916 the improvement was continuous and by 1955 the glaciers of the Alps had typically retreated about half a mile compared with their limits in 1865. The general warming coincided with the opening up of the American west, the heyday of imperialism, and a staggering growth in human populations which continues to this day even though the peak of the benign climate is past.

While the ice retreated in the Arctic and the mountains, contradictory trends in rainfall affected other parts of the world. The American Midwest dried out. That change in climate must take its share of the blame, along with human malpractice, for both the extermination of the bison ('buffalo') of the Great Plains and also for the disastrous Dust Bowl that made a mockery of American farming technology in the 1930s. In densely populated southern Asia, by contrast, the summer rains brought by more vigorous winds became more reliable for a time. The USSR also benefited from greater moisture penetrating its vast continent.

No precise date can be given for the ending of the recent 'climatic optimum'. In north-west Europe the first of the really cold winters came in 1939-40 but the last of the really good summers was in 1949. In Greenland the trend to cooler summers started earlier, in central Europe later. The oceans were still warming – catching up, as it were – in the 1950s. But 1950 is probably as good a

date as any for the end of the general upward trend in temperature which had begun in the 1780s. The winter of 1962-3 in England was the most severe since 1740 and showed what the 'new weather' could do.

The human species has been caught out by this climatic reversal in all sorts of major and minor ways. Climatic maps and tables based on conditions in the earlier part of the 20th century now turn out to refer to a completely abnormal period. The Danish government in the 1960s invested heavily in the Greenland fishing industry, only to have its plans confounded when the cod that had moved north into Greenland waters earlier in the century thought better of it and sheered off south again. Similarly, drought in Central Asia during the early 1960s frustrated Nikita Khrushchev's 'virgin lands' programme and created a dust bowl similar to that of the American Midwest in the 1930s. The chief cause for anxiety, though, about the present change of climate is the weakening of the Asian and African summer monsoons, with their threat of imminent famine.

This account of past climate has already exposed some tricky links in the world's weather machine. They can make a change that is harmful in one region helpful in another, and they connect the ice in the north to the rainfall elsewhere. A broader look at the weather machine itself is therefore appropriate.

The global winds

Christopher Columbus did not just discover America; he found the trade winds that blew him there. For most of his Atlantic crossing from the Canaries to the Bahamas his flotilla bowled along at a hundred miles a day with a steady wind from the north-east filling the sails. Columbus was lucky. The trade winds had edged unusually far north that year, and only for ten days or so did the ships encounter the light, variable winds more

typical of the route he had chosen for his crossing. Later sailors were to include it in the so-called horse latitudes, notorious for long and often fatal delays under a blistering sun.

As the navigators ventured all over the world's oceans their logbooks disclosed a pattern of prevailing winds. The trade winds slant from the east towards the equator, across both the Atlantic and Pacific Oceans and from both north and south of the equator. The prehistoric navigators of the Pacific must have known them well, long before Magellan. Reliable enough, in the days of sail, for the founding of empires, the trade winds are the most conspicuous feature of the pattern of winds at the Earth's surface. Along the equator itself are the doldrums, noted for their calms and heavy rains. Beyond the zones of the trade winds, in the southern as well as the northern hemisphere, are those horse latitudes — calm and cloud-free.

In the stormy zones further to the north or south, the 'prevailing' winds are less settled, especially in the northern hemisphere. Over the North Pacific and North Atlantic, the wind can blow from any point of the compass. There is, though, a bias in favour of winds from the west or south-west. The equivalent winds (west or north-west) are much plainer in the Roaring Forties and Screaming Fifties of the southern hemisphere. Around the poles winds are generally light and variable, but easterly winds are most common.

Weathermen call this pattern 'the general circulation of the atmosphere'. Although the phrase is explicit enough, it is somewhat cumbersome. I shall refer instead to 'the global winds', meaning those winds that are due to the world-wide operations of the airy part of the weather machine, in contrast with local winds such as sea breezes, thunderstorm squalls and the winds that blow up and down mountain slopes. The global winds are the key to understanding the extraordinary vari-

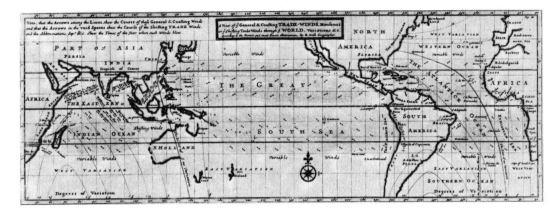

eties of weather and climate that occur in the different zones of the world, by land as well as by sea. If you see where you fit into the global winds the weather experienced in your own district makes more sense.

Savants ashore in various countries made many attempts to explain the wind reports of the sailing ships returning home. Some of their guesses were good, but they lacked vital knowledge that only high-flying aircraft and balloons could supply. Something close to a complete explanation of the global winds has emerged in the past few years from meteorological research of the new style. But one of the earliest accounts contained several of the crucial ideas and provides a convenient introduction to the modern version. It was an outstanding achievement in man's efforts to understand the world he inhabits.

In 1735 George Hadley, a lawyer living in London, reasoned as follows. Close to the equator, warm air rises; in colder regions, the air sinks. To keep these movements going, air has to flow towards the equator on the surface, and return from the equator high above the surface, eventually cooling and sinking. But the Earth is spinning and it is also ball-shaped. An object at the equator, whether a mountain or a piece of apparently still air, moves eastwards at a high speed (579 knots). An object nearer to one pole or the other has less far to travel around the Earth's axis, so it moves more slowly. Hadley reasoned that air heading south towards the equator starts off with too little eastward motion to keep pace with the spinning Earth nearer the equator. It lags behind, and slants in towards the equator from the north-east – the trade winds. But the air travelling north (or south) from the equator high above the Earth starts off with so much eastward motion that it overtakes the Earth and by the time it is half way to the pole it has become a strong wind blowing from west to east – eventually sinking to become the prevailing westerlies

of the Earth's stormy zone. Hadley also noted that the easterly and westerly winds had to be in overall balance at the Earth's surface, otherwise they would tend to make the Earth spin faster or more slowly.

Spotting some of the consequences of the Earth's rotation was one of Hadley's triumphs. He was wrong about them in detail. He did not know that the rotation deflects even an easterly or westerly wind to the right (in the northern hemisphere) and he underestimated by half the deflection of the northerly and southerly winds. What is more important is that Hadley's scheme for the global winds is incomplete. The flow of air towards the equator in the trade winds, and the return flow at high altitudes, operates, at most, only a third of the way towards the poles. And Hadley, like most people pondering the global winds right up to the present, was overenchanted with the steadiness of the trade winds and with the obvious need for the air to transport heat from warm regions to cold. This encouraged him to think primarily of air rolling steadily from the tropics to the poles.

The modern account of the global winds pays much more attention to the daily and weekly variations of the weather around the Earth. Instead of flowing steadily, the air is pulsating and the rolling over guessed at by Hadley is less important in most places than the horizontal whirls of storms. The new version derives from the findings and ideas of many weathermen over two centuries, but latterly from the work of the theorists at the Massachusetts Institute of Technology in particular. Carl-Gustav Rossby, Victor Starr, Jule Charney and Edward Lorenz are some of their names. Backing it up are working 'models' of the atmosphere in computers which, by calculation, reproduce the main features of the global winds quite well. Although the computer men do not know exactly how they do it, their success suggests that they have overlooked no

A chart of the trade winds made in about 1730, and (right) George Hadley's early attempt to explain them as the result of a great overturn of air rising from the tropics. It was a good guess but the modern version of the global winds is shown below.

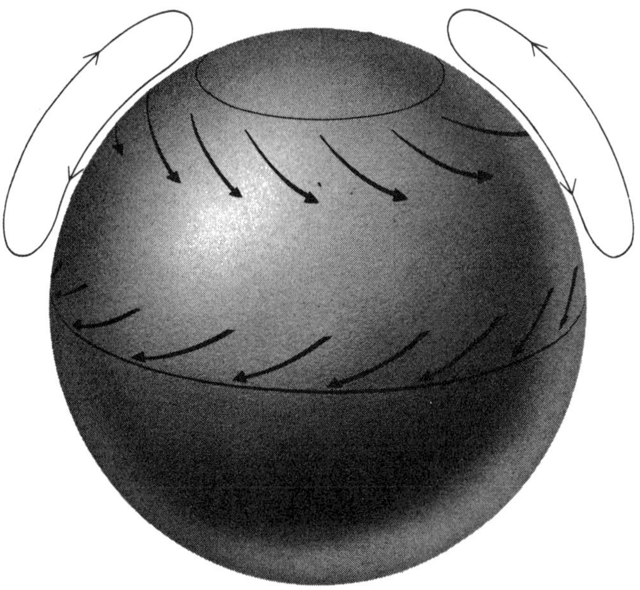

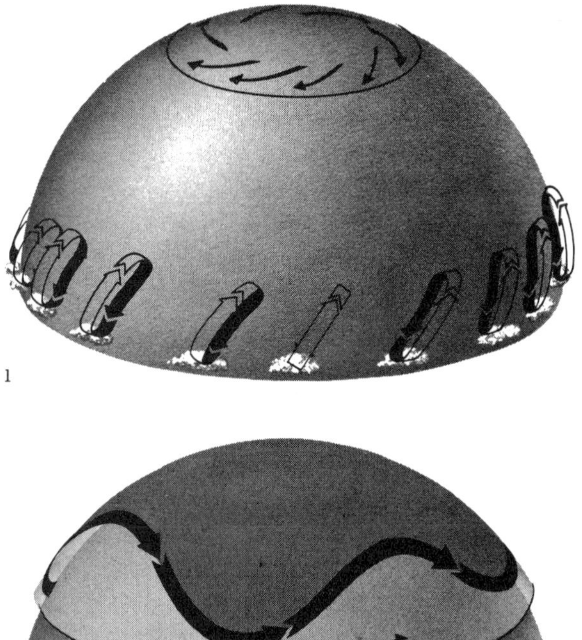

1

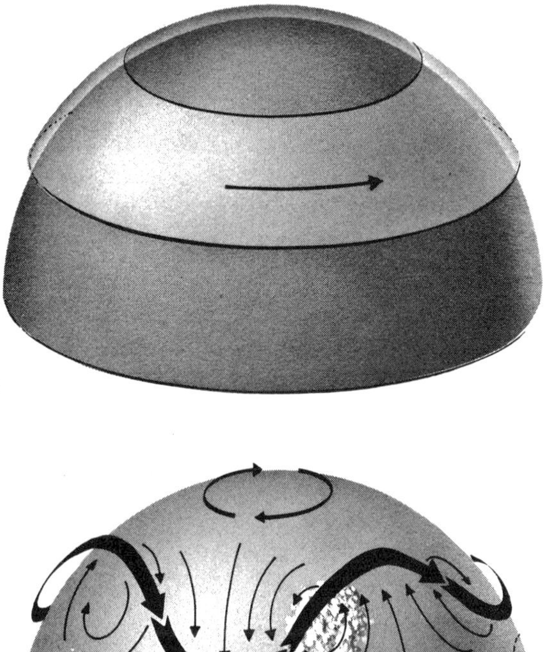

2

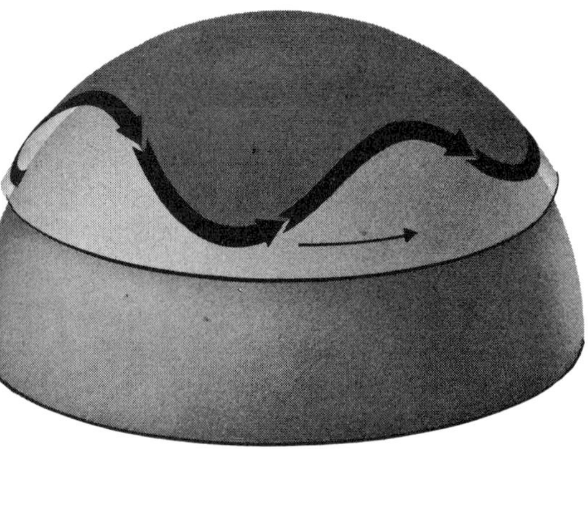

3

4

In the airy part of the weather machine, Hadley's system of overturning air really reaches only part of the way towards the pole (1). Cold air seeps out from the Arctic. Between these two zones, very strong winds (2) blow from the west in the upper air. At their northern edge is the zigzagging jet stream (3). In (4) the pieces are put together, with the further addition of the revolving winds at the surface beneath the jet stream in the stormy zone – the fair-weather 'highs' and the stormy 'lows'. The desert areas where the overturning tropical air descends drily are indicated by broken lines. (See also page 26.)

21

major factor and that the basic ingredients are represented in their mathematics.

Above all, the modern account is based on millions of measurements of conditions high over our heads, made by balloons, aircraft and satellites. One of the big discoveries, of the jet streams of the stormy zone, was made by aviators during the Second World War who, in new high-flying aircraft, found themselves being blown off course or making very fast flights eastwards and slow flights westwards.

Tropical thunderclouds draw energy from the warm oceans and from the moisture collected by the trade winds as they slant towards the equator. There will be more to say about how these clouds work in the next chapter. Essentially they wring moisture out of rising air, which releases energy and helps to raise the air even more. The tropical clouds pump air to a height of ten miles; some of it goes even higher, to supply the stratosphere, but that is another story. In the Atlantic and Pacific sectors of the Earth, the air that has stopped rising in the clouds spreads sideways both north and south from the warmest zone.

The whirliness communicated to the air by the spinning Earth is second only to the Sun itself in fixing the overall performance of the weather machine and the courses of the global winds. Much of it is simply a matter of the high air moving north or south from the tropics and deflecting eastwards, but the air also contains eddies of revolving air that reinforce the whirliness. Let us follow the air into the northern hemisphere; its actions in the southern hemisphere are similar but less well observed.

The whirliness of the high-altitude winds coming from the tropics accumulates in a broad river of air, rushing around the stormy zone from west to east – the 'upper westerlies' which contain the main jet stream. At this 'river', the air drifting north from the tropics is effectively halted. This is as far as the circulation envisaged by Hadley can reach and air from the tropics, running out of energy, has nowhere to go but down. It sinks most commonly in a zone running through the North Pacific from Japan to California, and across the Atlantic from Florida to Gibraltar. These are the horse latitudes, of clear skies and light winds, but the descent supplies the air needed for the trade winds blowing back to the equator. Hadley gave essentially the right reason for why the trade winds have a lot of east in them: with winds outside the tropics blowing from the west, the trade winds have to blow the other way if the winds are not to make the Earth spin faster.

Air rising in the tropics makes clouds and rain; air that is sinking is unable to make rain, because the air warms itself as it descends and its sponge-like capacity for moisture increases. The world's principal hot deserts lie in the zone of descending air on the poleward edge of the trade-wind zone: the US South-west, the Sahara and Arabia in the northern hemisphere; the Atacama, the Namib-Kalahari and Australia in the southern. In the Arctic and Antarctic, too, the air is often sinking over the frozen seas and lands. It seeps outwards across the surface in light and inconsistent winds. Relatively little moisture evaporates from the cold Arctic ocean or the polar ice, and snowfalls come mainly from comparatively warm, moist air guided far to the north by the jet stream.

The main function of the weather machine is to spread the abundant heat supplies of the tropical regions to cooler parts of the Earth. In winter the task is more demanding and the global winds are correspondingly more boisterous. The ocean currents play an important part, too, but the chief work of sharing the warmth from the tropics with the poles is done by the main jet stream and its attendant eddies. They create the stormy weather of the middle latitudes.

A pump for lifting air. Tall thunderclouds draw energy from the tropical oceans and help to power the weather machine.

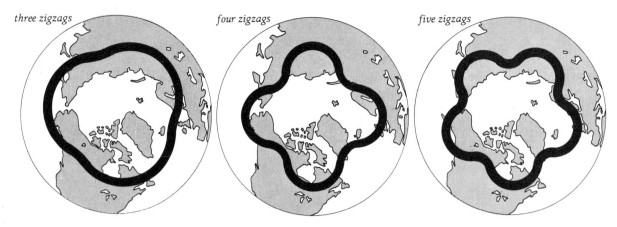

three zigzags *four zigzags* *five zigzags*

Zigzags in the upper air

The all-embracing whirl of the jet stream high overhead feeds on the lesser eddies of air. One of the strange properties of fluids on the rotating Earth enables the circulating wind to be reinforced by winds that blow in its direction and avoid being arrested by contrary winds. As a result strong winds blow from west to east around and around the world without respite, most vigorously at about six miles above the surface. In the core of the jet stream the wind speed varies, but it can reach more than 200 knots. Sixty knots is more typical.

Some weathermen prefer to confine the term 'jet stream' to the very strongest tubes of high wind which are not continuous around the world, although the current of the upper westerlies is continuous. To complicate matters, the atmosphere has a propensity for making jets: they occur also in the tropics and at low levels in the stormy zone, while the high-level stream in the stormy zone can split in two, with one jet skirting the desert zone and the other wandering much nearer to the pole. By *the* jet stream I mean, as many meteorologists also mean, the continuous mainstream on the poleward side of the upper westerlies blowing around the stormy zone. In the northern hemisphere it passes over North America, Europe and the USSR. The characteristic storms of the stormy zone march eastwards in accordance with the general flow.

A piece of air in the jet stream, or a balloon riding in it, takes about two weeks to circumnavigate the globe, but it does not take the quickest route. The jet stream zigzags. It carries warm air slantingly towards the polar regions and then turns, drawing cold air away from the pole, so helping to share the world's heat more equitably. At the Earth's surface, big eddies related to the jet stream, sometimes fixed and sometimes moving, also busily exchange tropical air for polar air.

The chief barrier that the jet stream encounters in the northern hemisphere is the chain of the Rocky Mountains rising up beneath it as it sweeps in from the Pacific Ocean over North America. Even the winds of the upper air have to climb a little to clear the obstacle. As they do so, the jet stream swings to the north, and then immediately to the south as it descends on the far side. The zone of strong winds has to flatten and broaden itself over the mountains; in the process its whirliness diminishes and it yields to pressures driving it northwards. Beyond the mountain ridge it recovers as it thickens again, which causes the ensuing swerve to the south. The result of behaviour over the mountains is to keep the Rockies covered most of the time with warm air at high altitudes – a semi-permanent feature of the world weather maps.

The path that the jet stream follows after leaving the Rockies behind it is then variable. Its zigzagging is not casual. It has to catch its own tail over the Rockies, by the time it has come around the Earth – otherwise it would create a vacuum somewhere. So it has to make a precise number of zigzags, typically four, often three or five, and sometimes as few as two or as many as six. In winter, when the air is moving most vigorously, three zigzags may be common. In summer there tend to be more. But because the choices are limited there are other places around the world, besides the Rockies, where the upper air is frequently warm due to a northward swerve of the jet stream bringing warmth from the south. In the three-zigzag pattern of a mild winter they often come over Iceland and over central Russia just east of the Ural Mountains. In between, the jet stream brings tongues of cold air southwards, over New England, over the Moscow region and over eastern Siberia.

There are two mechanisms for making the zigzag pattern and competition between them is one reason

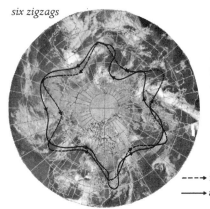

The jet stream adopts ever-changing patterns around the pole, but some are dominant in particular seasons or climatic periods. The diagrams are idealised. In the case of six zigzags, the ideal is compared with less tidy reality. Note the dense patches of cloud in the satellite picture, signifying storms, that occur near the parts of the jet stream trending towards the pole. The whole satellite picture of the northern hemisphere is reproduced below.

----→ ideal jet stream

——→ actual jet stream

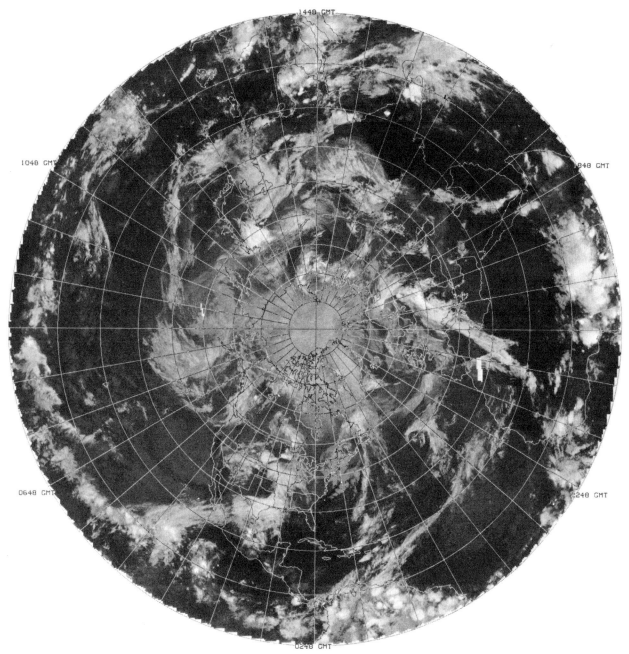

NOAA–2 *infra-red mosaic, night of 5-6 May 1973.*

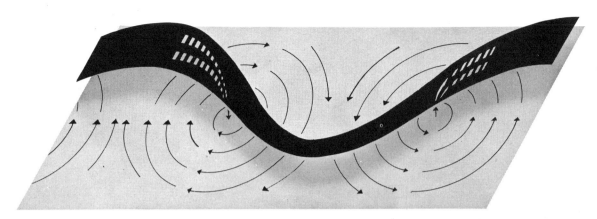

why the weather in the stormy zone is so variable. One mechanism is geographical. Apart from the Rockies, other features of the ground and the sea exert their influence, notably the channel between Greenland and Scandinavia that helps funnel cold polar air over Iceland. There is an obvious conflict with the three-zigzag situation just mentioned, which warms Iceland. Similar channels of cold air, made by low-lying ground, occur in three places in Russia: east and west of the Urals and in the valley of the Lena River in eastern Siberia. The temperature of the sea, especially when it is much warmer or cooler than nearby land, also helps to fix the pattern – the route for the jet stream. So does ice, whether lying on the land or floating on the sea. These influences help to fix our weather.

But the jet stream has a will of its own. Like a ship that yaws uncontrollably from side to side, it is unstable. Once it starts to swerve, effects of the Earth's rotation encourage it in its waywardness, though a strong jet stream zigzags less often than a weak one. Also its obligation to catch its own tail and its need to make a definite number of zigzags around the Earth have ultimately to override any other factor. So the pattern of zigzags is a temporary compromise between the dictates of conditions at the Earth's surface and the jet stream's own choice of a route; the pattern can change every few days.

All this about the jet stream and its path concerns the upper air. Warm or cold air aloft does not directly determine conditions at the surface, which is just as well because 'warm' and 'cold' are relative terms. All the upper air is well below the freezing point and the coldest place on Earth is twelve miles above the equator, where the air temperature is −80 degrees. But there are very important indirect effects of the upper air. In particular the zigzags of the jet stream help to fix the positions of storms and fine weather in the lower air. Around the

stormy zone at any moment there is always a fairly regular alternation of regions of fine and wet weather, matching the zigzags of the jet stream.

Where the jet stream zigs towards the tropics, the air tends to concentrate and press downwards, creating an area of sinking air and fine weather at the surface (a 'high' or anticyclone). But where it zags towards the Arctic, the jet-stream air spreads out and exerts an upwards suction. At the surface, the results are rising air and wet and windy weather (a 'low' or depression). The persistent 'high' near the Rockies and a persistent 'low' in Iceland (which corresponds to a four-zigzag situation) are conspicuous features of the present climate of the northern hemisphere.

There is give-and-take. The jet stream helps to create conditions at the surface, but it has to adjust its course to conditions imposed from the surface, perhaps as a result of a warm sea or cold ice created by its own earlier activities. The actions and reactions of the jet stream are thus important factors in changes of climate. In summer it moves north with the sunshine and becomes weaker because the contrast between tropics and poles is much less in summer. Over North America at the beginning of May, it skips from a typical winter position over Oklahoma to the Canadian border. The spells of fine weather often experienced in north-west Europe in May are due largely to the global winds becoming locked into a particular pattern. A dry area of descending air and high pressure persisting for some days or weeks over Scandinavia bars the way to the procession of Atlantic storms. Weathermen call it a 'blocking anticyclone'; I shall use the briefer term 'block'. In early summer the block over Scandinavia often has benign results in north-west Europe, but blocks occurring at other times and places can be disastrous. Whenever you hear of heat waves or record snowfalls or of droughts or floods, in the stormy zone,

it is reasonable to suspect that a block has interrupted the normal flow of weather from west to east.

The jet stream is like a meandering river in spate. If its zigzag path has become particularly kinked it may suddenly take a short cut – just as a river can break through to a lower part of its course, leaving a lake to one side. When the jet stream acts in this way it can cut off a huge pool of warm air on the polar side of its new course. The warm pool is now 'out of the wind' as far as the jet stream is concerned and it is almost stationary. Thus a block is built which can prevent the jet stream re-entering the area for a week or more. Blocks of a different kind crop up in a similar way on the tropical side of the jet stream; they are stationary depressions that modulate the normal trade-wind weather in islands on the edge of the tropics, like Hawaii and Madeira, but they are smaller and less important than the 'blocking anticyclones' to the north.

Blocks are most likely in spring and autumn, when the jet stream has to change the number of its zigzags to suit summer or winter conditions and when the land is warming or cooling rapidly. Blocks in summer are serious if they bring drought or floods during the farmers' growing season, and in winter if they cause long periods of unusual snow or cold winds. England seldom has a white Christmas, even if there has been snow in the weeks before. The jet stream over the Atlantic is at its strongest in late December; it blows away any blocks that might be sending cold air south, and brings in warmer weather from the ocean.

We still have to account for the mobile storms, the travelling depressions, that dominate the weather in the stormy zone. The storms drift eastwards, taking half a week, for example, to cross the Atlantic from Newfoundland to Europe. They travel more slowly than the air in the jet stream, but faster than any

changes in the main pattern of zigzags in the jet stream's path. The jet stream's habit of wriggling, and so creating areas of upward suction that correspond to storms at the surface, continues even when its main path is settled. It thus encourages relatively small storms to form – about 1000 miles in diameter compared with about twice that size for the persistent depressions and 'highs' produced by the main zigzags. The air in the storm is whirling around and around, but the whole eddy drifts eastwards under the jet stream. The corresponding wriggle in the jet stream itself is like a tremor passing eastwards through the main zigzag pattern.

Again, the jet stream does not have things all its own way; surface conditions have to be right and in practice the storms originate most frequently to the east of mountain barriers (notably the Rockies) and just off the east coasts of North America and Asia. The Atlantic south of Newfoundland is the source of much of Europe's weather.

Warding off the cold

The next time you are seasick, or spend a holiday looking out of the window at pouring rain, be comforted that it is all in a good cause. The depressions that are so

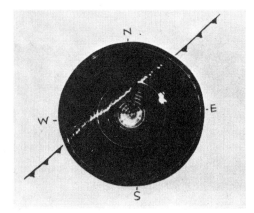
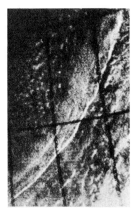

A low-level jet of warm air streaking into a depression defines the stormy 'cold front', which sometimes shows up clearly in radar displays and satellite pictures (right). Near the centre of the depression the incoming warm air swerves and rises. As shown in the diagram, it makes further fronts at different levels in the air, producing bands of rainfall. (After K. Browning.)

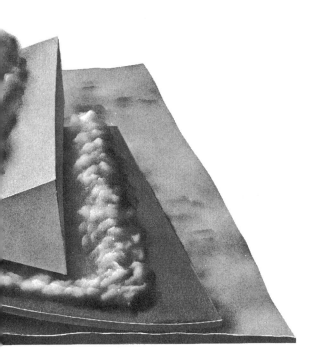

familiar a feature of life in most parts of Europe and North America are eddies in the global winds busily transferring heat from the tropics to the polar regions and so helping to ward off the next ice age. They are also the chief suppliers of rainfall in the stormy zone, but the weather actually experienced, as the 'warm front' and 'cold front' of a depression, is in many respects incidental to the main work of the global winds.

Older theories described depressions as products of the encounter of cold polar air and warm tropical air along the fronts that are so conspicuous on the weather maps. The new account of the global winds makes no appeal to the idea of a front, and explains depressions as inevitable eddies in the west-to-east airflow around the stormy zone. The fronts turn out to be by-products of a depression, rather than its cause. Yet fronts are real enough and they engineer the stormiest weather.

A large depression contains a sector of warm air lying (in the northern hemisphere) roughly to the southeast of the storm centre. It is the wettest part of the storm, where the air rises most vigorously. As the storm moves eastwards, the warm front passes first, bringing the warm sector to you, and then as the warm sector leaves you behind, perhaps a day later, the cold front comes along.

The warm sector is wider in the upper air than at the surface, because the warm air overrides the cold air at both fronts. Gliding upwards over the cold air, as if up a hill, the warm air itself cools and forms clouds and rain. Towards the centre of the storm the two fronts come together and squeeze the warm air completely off the surface. The warm air spills a little way across the centre, causing clouds and rain on the far side, but otherwise the cold sector of the storm is generally fair. That leaves the warm sector of a depression as the chief rain factory for the stormy zone.

The western part of the warm sector, just ahead of the cold front, is also the scene of the most violent conditions experienced in the stormy zone. The general flow of air creates strong, changeable winds and prolonged rain, but it also spawns severe thunderstorms and tornadoes. While the warm front is a vague region of high cloud gradually coming closer to the surface as the front approaches the watcher, the cold front is a sharp line drawn across the weather. Any sharp line in the air requires a vigorous explanation, and that leads to a more vivid picture of the rain factory. Here I follow very recent descriptions by Keith Browning of the British Meteorological Office, who has himself carried

out many of the key observations of fronts, chiefly with radars that watch belts of rain clouds crossing the British Isles.

The warm sector is fed by a jet of warm, moist air, sucked into the depression from the south-west. This jet flows only a thousand feet or so above the surface, far lower than the main jet stream that encircles the world. It is about 100 miles wide and the wind is strongest on the extreme left-hand edge of the jet (in the northern hemisphere), where it reaches 50-60 knots. Immediately to the left of the maximum wind is the cold front, and it is the jet that defines the front and keeps it sharp. But the jet is also corkscrewing, dragging air upwards on its left and pushing it down on the right. The rising air on the left, precisely at the cold front, forms a line of clouds and heavy rain.

The low-level jet curves gradually northwards towards the centre of the depression, where it makes an abrupt turn to the right, sweeping outwards, but less vigorously, along the warm front. It rises over the cold air outside the warm sector and makes the characteristic clouds and rain that run ahead of the warm front. Riding on this 'conveyor belt' the warm, moist air meets the cold and dry air of the main jet stream above. The encounter makes new cold fronts, high above the surface, each complete with a strong wind along the front and a band of clouds giving heavy rain. That is why people on the ground so often experience, as a depression passes, a succession of periods of heavy rain, culminating in the arrival of the cold front at the surface. But the rainband along a front is not always continuous; a low-level jet stream, like its big brother in the upper air, wriggles in a way that encourages clouds to form at some places and discourages them in patches in between.

A word about tornadoes. The thunderstorms of the cold front of a depression often spawn these small but

most violent of tempests in the warmer parts of the stormy zone. They are commoner in England, the Netherlands and Germany than is generally realised, but they are notoriously an American phenomenon. After 1973 had been recorded as the United States' worst year ever for tornadoes, with more than 1000 of them, a monstrous regiment of 90 tornadoes assailed ten eastern states on 4 April 1974, with much loss of life. A recent discovery is that families of severe tornadoes are generated by colliding squalls of cold air from nearby thunderclouds. The core of air thus set spinning sucks in moist air from a wide area, swirling faster and faster, just like water down a plug-hole. The result is a visible vortex of strong rotation and suction, long and narrow, dipping across the countryside.

Just how tornadoes sustain their terrible force remains a mystery but, to imitate their action, Theodore Fujita operates a machine in his laboratory at the University of Chicago. Revolving sets of cups force the air to spin while a suction system draws air upwards. With this equipment, Fujita finds that too much suction makes the model tornado collapse, and what actually keeps the vortex going is a tube of air coming down at very high speed around the outside of the visible vortex. In real storms, Fujita has seen clouds growing unusually tall and then their tops dropping like a hammer; he thinks that the fierce down-draught so produced is the trigger for a tornado.

The rising air of the depressions and their fronts, and the intervening times of descending air and blue skies, give the weather of the stormy zone its special fascination. In the course of a year, any window in western Europe, for example, or Japan or the northern part of the United States, looks out upon as rich a variety of weather as a reasonable person could wish for in a lifetime – a touch of the North Pole and of the tropics, a taste of desert drought and a soupçon of monsoon.

Snowfalls whitewash roofs and grass as if nature were tired of the colour scheme. Hoar frost makes filigrees of commonplace objects – a spider's web or a street lamp. Most mocking is the fog that robs us of our sight, leaving a man to grope with his foot for the kerbstone, and great ships to moan like lost children. One fog that spread from Wales to Finland in November 1948 brought half Europe to a halt for a week.

High summer, when the hot road scalds the air and makes mirages, and the grapes ripen under the solar grill, is a time for love – as recorded in the peak in the human birth rate the following spring. But when smart little cumulus schooners are drifting across the summer sky, and the air is close, watch out for the full-rigged thunderstorms that lay themselves alongside defenceless fields and pound them with electric fire and hail.

Contrast two characteristic kinds of days in spring or autumn. On one, the air is descending. During the night, the ground has radiated its warmth into the starry night. The morning grass is awash with dew and there is mist in unventilated pockets of the landscape. All day the Sun shines kindly, neither too hot nor too cold, and by the time it sets the sky is rosy with newly raised dust. If the other kind of morning is anything but grey, the sky is deep red on account of all the water droplets between you and the Sun. You are in or near the warm sector of a depression and several times during the day the heavens open, as the cold fronts in the upper air and eventually at the surface pass by like a stately procession of road sprinklers.

A change in the frequencies of those two kinds of days can amount to a substantial change in climate. In fact moderate changes of climate do not involve weather that is especially strange, but only 'too much' or 'too little' of certain normal weather conditions, or occurrences at the 'wrong' time of year. For example, a storm that is warming in winter can be cooling in sum-

The explorer James Cook met waterspouts off New Zealand. They are akin to tornadoes on land, but less energetic. (By W. Hodges.)

A disastrous tornado at Union City, Oklahoma, on 24 May 1973. No previous storm was ever so closely observed – movie cameras, still cameras, radars, weather stations and a satellite all provided information. These photographs show the vortex forming, 'touching down' and leaving its trail of damage.

mer; a region of descending air that brings blazing heat in summer may make a winter bitterly cold. And because of the alternations of wet and dry areas, beneath the zigzagging jet stream, a worsening of one country's weather may be associated with benefits to another.

The machine changes gear

The only way to make sense of apparent contradictions in climatic changes in different parts of the world is to see the weather machine working as a whole, and in particular to pay attention to the patterns of global winds. The first scientist to emphasise this strategy for obtaining a deeper understanding of climatic changes from one century to another was Hubert Lamb, a scholarly man who spent many years in the British Meteorological Office gradually piecing together scraps of information about the weather in the past. Together they give a lively account of changes since the end of the most recent ice age, and especially during the past ten centuries. Lamb now directs the University of East Anglia's climatic research unit. He is also chairman of the World Meteorological Organization's working group on climatic fluctuations; his knowledge on the subject is encyclopedic.

During the past 200 years in Europe measurements of air pressure were made with barometers at a sufficient variety of places to allow the modern investigator to discern regional patterns of the wind and weather. Measurements of air temperature also exist in old manuscripts and books. Continuous records of temperatures in central England go back to 1680, and in the eastern United States back to 1738. There are 250 years of rainfall statistics for the Netherlands, and a little more for London. Some of the most valuable compilations give the temperatures of the sea surface of the North and South Atlantic as measured by British and American

ships in the period from 1780 onwards. They show that in the early 19th century, during the Little Ice Age, the great warm current of the Gulf Stream took a more southerly course across the Atlantic than it does now and it swung further to the south as it approached the coast of Europe. Instead of warming Britain and Scandinavia as it does today, it benefited Spain. The South Atlantic was generally warmer at that time than it is today, but it did not supply as much warm water to the North Atlantic. While the Arctic ice was more extensive, the Antarctic ice was confined more closely to the southern continent.

Lamb took the story further by reconstructing maps of typical air pressures over the North Atlantic and Europe for the late 18th and early 19th centuries, for January and July. To the few observations available he added patterns of high and low pressure (descending and rising air) that would produce the winds and weather reported at various places. The maps showed that, in the winters around 1800, the winds coming into western Europe from the Atlantic were weaker than in the winters of the first 40 years of this century. A weaker flow of air from the ocean would have made western Europe more vulnerable, in winter, to frosty air spilling out from Scandinavia and northern Russia.

The want of vigour in the warm Atlantic winds may also help to explain the weaker behaviour of the Gulf Stream and the extent of the sea ice spreading from the Arctic. In the first decades of the 19th century the ice virtually encircled Iceland. There was ice on the coast for an average of nine weeks a year at that time, compared with about ten days in the period 1920-50. (The ice lingered even longer in the 1690s, the 1780s, the 1860s and the 1880s.) But the weather machine can change in more striking ways than by enfeeblement.

From old accounts of weather experienced in Europe, since long before the start of scientific measurements,

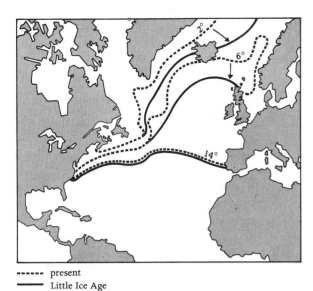

Peculiarities of the weather (right) 200 years ago and half-way around the world. The areas of abnormally high (light) and low (dark) pressure were deduced from reasoning based on the patterns of growth in American trees during the period in question, as mapped on page 36. (After H. C. Fritts.)

----- present
——— Little Ice Age

Some changes in the weather machine during the Little Ice Age. This map shows the southward shift of the warm Atlantic water. The Gulf Stream, which sweeps across the Atlantic and delivers warm water to Europe, was less effective. (For a more drastic shift in the Gulf Stream, see page 115.)

An associated change is shown in the lower map – the expansion of the area of the North Atlantic covered by sea ice in April, during the Little Ice Age. (Both maps after H. H. Lamb.)

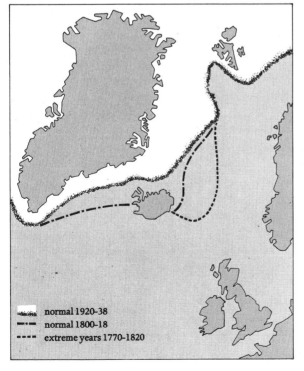

████ normal 1920-38
—·—· normal 1800-18
----- extreme years 1770-1820

Lamb has deduced the tracks of storms for periods when the climate differed markedly from our own. The 11th century AD was an exceptionally warm period in the Viking era. At that time summer storms tracked eastwards across the Atlantic into Europe most frequently in the latitude of Iceland and northern Norway. In the latter half of the 16th century, during the Little Ice Age, conditions were completely different. The Atlantic summer storms were coming in over Scotland and Denmark, with quite a few of them entering the Mediterranean – a rare occurrence nowadays. It was as if the countries of western Europe had moved 300 miles to the north. More precisely, the pattern of global winds, including the jet stream, had shifted south.

The weather could also shift eastwards or westwards. For example, around the 1740s Britain was experiencing relatively dry summers while the Moscow region was repeatedly wet; three decades later, the situation was reversed. For the explanation, we have to appeal to the wavy motion of the jet stream, blowing from west to east over the stormy zone but also zigzagging north and south. The typical number of zigzags helps to fix the positions of wet and dry areas around the world. These positions, remember, are a compromise between the attempts of the jet stream to keep to a consistent wavy pattern right around the world, and the mountains and plains, oceans and ice-packs that impose their own pattern.

In the winters of the warm early part of our own century, the number of jet-stream zigzags was predominantly three. With the pattern anchored to the Rocky Mountains, three zigzags put western Europe in an area where warm air was funnelled northwards from the tropics, even in winter. But in the Little Ice Age considered by Lamb, four zigzags were probably more common because the jet stream was weaker and travelling further south. If so, the area of warm souther-

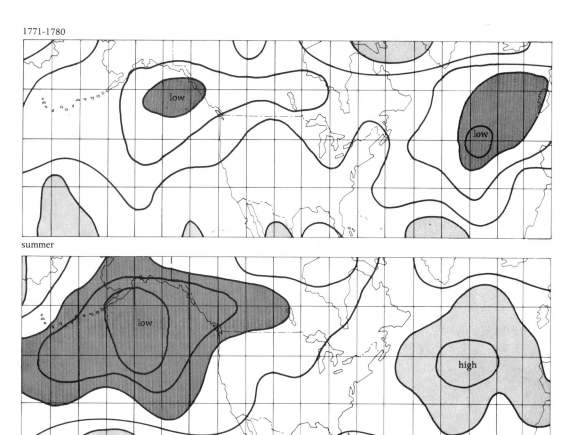

summer

winter

ly winds would have been pulled westwards into the Atlantic, while Europe suffered wintry blasts from the Arctic. Such patterns of east-west variation, linked to the jet stream, also explain why the worst winters recorded in Moscow and in London seldom coincided. In the 1780s the weather of May and September was much poorer in north-west Europe than it is now. The blocks over Scandinavia, which create the fine weather in those months nowadays, were forming instead over Greenland, as they tend to do now in April and October. There is something charming about Lamb's ability to reconstruct the workings of the global weather machine over the heads of our forefathers, all unbeknown to them. But life was grim in the Little Ice Age – and not only for the north Europeans.

The trees remember

Long before white settlers harried them from the homelands, the Pueblo Indians of what are now the southwestern United States were forced into great migrations by persistent drought. There, human beings have contended not with cold but with a variably arid climate. One episode was in AD 1300, following an excep-

tionally dry period starting in 1271, when the world's glaciers were just beginning to advance again after the warmth of the Viking era. The farming civilisation of the Anasazi was at its peak, but they had to abandon their sacred mountains and cultural centres at Mesa Verde (Colorado), Chaco Canyon (New Mexico) and Kayenta (Arizona). Among the buildings they deserted were three great, newly completed cliff-dwellings in what is now the Navajo National Monument. Two centuries later, when the Little Ice Age was beginning to grip Europe, redoubled drought made some tribes move again. In both episodes, as prayers and rainmaking ceremonies failed in their purpose, people were forced to leave the mountains and settle by the rivers for a more reliable water supply.

The old trees in the region do not forget the ruination of the land by drought. By counting the rings of annual growth preserved in a tree-trunk you can in principle tell its age; in practice, more elaborate dating is done. By seeing how the thickness of the rings varies you can tell whether the tree fared well or badly in a particular year. Nor is that all: by recognising sequences – runs of good or bad or mixed years – in timbers long-since hewn, you can match them to corresponding patterns

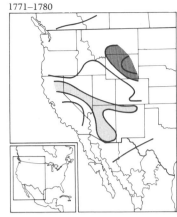

in surviving trees. Scientists at the Laboratory of Tree-Ring Research at the University of Arizona exploit the great longevity of trees in the arid south-west not only to date local archaeological sites but also to help correct the radiocarbon time-scale used by archaeologists all over the world. Andrew Douglass, the American founder of tree-ring research, was an astronomer who first turned to trees in 1901 looking for signs of the Sun's weather – the sunspot cycle – affecting the weather on Earth. But only recently has the promise of tree rings begun adequately to fulfil itself, as far as past changes of climate are concerned. Much of the credit belongs to Harold Fritts, also at the Laboratory of Tree-Ring Research. Fritts is a plant ecologist whose interest in weather and climate led him to join the Arizona laboratory in 1960. He uses computers and weather maps to resolve problems that confounded the pioneers.

In principle it looks easy enough. A pencil-thin 'core' bored from a tree's trunk (which scarcely harms the tree) shows the varying thicknesses of annual growth. Reckoning backwards reveals that the ring for 1776, for example, is narrow: the tree grew poorly, so the weather was bad that year at the place where it stands. The difficulty comes in working out what 'bad weather' (or 'good weather' in other years) really meant. Sunshine, warmth and rainfall are three quite different factors that affect the growth of a tree. They are only the most obvious ones.

Trees placed differently in the landscape, on a windy crest or in a hollow, by a stream or in a forest, will be vulnerable to quite different effects of weather. A tree's height above sea level will affect its response to changing circumstances. In a hot climate low summer temperatures may benefit the trees, while for those at the fringe of the Arctic the warmer the better. Hangover effects operate from one year to the next; a very bad year can cripple a tree even for the following good sum-

mer. Nor is climate alone responsible for all the variations. Competition with other trees for light and moisture, assaults of diseases and pests (which may themselves be governed by the weather) and the internal machinery of life within the tree all influence the pattern of growth. A forest fire can alter the tree's environment for better or worse. So great care is needed in drawing conclusions about climatic changes from the thicknesses of the rings of growth.

Computers have transformed tree-ring research. They can take in information about thousands of trees, discount flukey events in individual trees, and detect consistencies or contrasts between tree growth in different locations. A mass of evidence for different species of trees shows that they respond differently to variations in temperature and rainfall at different seasons of the year, so one species can be played off against another to eliminate uncertainties. Deducing past climates from tree rings becomes an exact science. But of course there have to be plenty of samples, and you still have to go repeatedly into the field, choosing each tree carefully before taking a core.

Fritts and his colleagues have done most of their work for the western United States and adjoining parts of Mexico and Canada. Much as Hubert Lamb has deduced Atlantic storm tracks from European weather history, they have produced a series of weather maps, interpreting the climatic variations in terms of the global winds. Going back to 1700, when Spaniards ruled in Santa Fé, the maps extend from the west of North America right across the Pacific and the Atlantic. They show, for each period, patterns in the global winds needed to account logically for the distribution of rainfall in the west of North America, when trees in one area may have been enjoying ample rain that was denied to other trees a few hundred miles away.

Now Fritts is extending his studies to the eastern

Harold Fritts (far left) bores into a tree to obtain a sample of its rings of annual growth. From statistics on many trees in the west of North America, geographical patterns of good and poor growth emerge, for each selected period in the past. A typical map shows regions in western North America where the ring widths were wider (dark tint) and narrower (light tint) than usual; the other lines mark the strips of territory where ring-widths were normal. The trees that give the longest record of climatic change are the bristlecone pines (below), some of which have lived more than 4000 years.

To bring water to the rice fields of Asia the weather machine has to contrive the great overturn of the monsoon.

United States, with tree-ring samples from more than two dozen sites. He is also urging tree-ring experts in the USSR and various European countries to pool their information and to use the same computer programs for analysis. In that way they might extract the maximum comparable information about past climates from the long-playing records of the trees, all around the northern world.

While Fritts has striven for comparisons of tree rings over wide areas for a few centuries, a colleague, Valmore LaMarche, has been extending the climatic record of tree rings as far back as possible. He uses the bristle-cone pines of the White Mountains of California. These trees survive for more than 4000 years and matching their rings with remains of dead trees enabled the Laboratory of Tree-Ring Research to push the dating record back more than 8000 years. That work was begun with trees living near the lower border of the forests where the climate becomes too dry. LaMarche has now compared with this record of the lower trees the annual growth rings of dead and living bristle-cone pines near their upper limits on the mountainsides, where the climate becomes too cold. While the lower trees grow well or poorly according to rainfall, the upper ones want warm summers.

With these remarkable trees at the upper-tree line, LaMarche can see the changes in summer temperatures going back to 3435 BC. The record generally corresponds with the world story of climatic change given earlier, though with some interesting variations. It shows cool summers in the White Mountains around 3000 BC, 1300-200 BC, AD 400-1000 and 1320-1850 (the Little Ice Age). A very warm period shows up from 1050 to 1320 and again in the early part of the 20th century.

But the most striking conclusions come from comparisons with the bristle-cone pines at the lower, arid border over the past 12 centuries. LaMarche is able to categorise the various periods by moisture as well as by warmth. In the White Mountains the terrible drought that drove the Anasazi from their homes in 1300 shows up clearly, but we also learn that the climate was warm. The early part of the Little Ice Age (1320-1650) was cool and wet there; the latter part (1650-1850) was cool and dry. In the one period the jet stream over the Rockies was displaced southwards, in the other northwards, but in both parts of the Little Ice Age the global winds were evidently weak. Most strikingly the 'cool and wet' pattern matches very well with Hubert Lamb's deduction, based on European information for the summers of 1550-1600, of a jet stream with five zigzags, one of which brought it southwards at the American west coast. The early decades of this century, like the period around 1100, were warm and moist, a sign of vigorous global winds. Recently the White Mountains have become markedly dryer.

Human beings need water so badly that the map of world population looks very like the map of world rainfall. The resemblance is even closer when you make allowance for those who live by the great rivers rather than on the wet mountains that supply them. Warming and cooling in the northern lands are linked, via the weather machine, with changes in rainfall in other parts of the world. One cannot generalise about the connections: cooling brings less rainfall in some places and more in others. Even in one locality, such as the White Mountains, the outcome of the cooling can change from one period to another, depending on which way the global winds blow, and how strongly.

Recurrent winter and spring droughts in northern Europe and the Russian grainlands in the early 1970s are related to a weak jet stream and the tendency for the weather to travel northwards or southwards, rather than from the west, off the ocean. In 1974, Norway had

the driest spring ever recorded. The North American grainlands suffered devastating floods and then drought later. Drought is the greater risk when the world is cooling, as at present. Conversely, when the Earth was very much warmer, around 45 centuries ago, the rain reached into areas that today are dry.

Rain for half mankind

The Harrapan was one of the great civilisations of the ancient world. It flourished for about six centuries, cultivating grain, cotton, melons and dates in the Indus Valley, in what is now Pakistan and north-west India. Here farmers had been busy for thousands of years before, but the prosperous artisan cities, Harrapa and several others, were the climax of the achievements of the people of the Indus. Around 1800 BC they went into a decline and eventually succumbed to Aryan invaders from the north-west.

Since the 1930s archaeologists have warmly debated whether variations in the climate were responsible for the changing fortunes of the Harrapan civilisation. The weight of professional opinion came down against any such idea, chiefly on the grounds that the farmers relied on the rivers coming from the Himalayas, rather than on local rain. In 1971, the whole issue was reopened. Discoveries of fossil pollen showed that an area in north-west India at the edge of the Harrapan region, which is now arid, was formerly a land of rich vegetation.

An expert in fossil pollen, Gurdip Singh from the Institute of Palaeobotany in Lucknow, investigated salt lakes in north-west India and found that they were formerly fresh-water lakes in the midst of richly vegetated land. The most interesting of the lakes is Lunkaransar, near Bikaner, deep in the Great Sand Desert of Rajasthan. Here, today, the hot, moist wind of the summer monsoon delivers scarcely any rain; instead it piles up drifts of sand and the dunes march slowly across the countryside. The vegetation is sparse. But dig just a few feet through the salt of Lunkaransar and you come to neat layers of mud, laid down when the lake carried fresh water four thousand years ago. And in the layers Singh found pollen of bulrushes and sedges. The lake also collected, from the surrounding land, pollen of grass, jamun trees, mimosa and many other species. Jamun trees need at least 20 inches of rainfall a year. There are also traces of cereals and charred remains of plants: man was evidently farming in what is now desert and in fact those traces go back 9000 years or so, to the very earliest agriculture. Underneath are the sand dunes of a fossil desert; it corresponds to the last phase of the most recent ice age, when the area was as arid as it is now. This fossil desert, incidentally, is one of many pieces of evidence that now sweep into the dustbin the long-standing theory that ice ages in the north necessarily meant abundant rain in the tropics.

Judged by the nature and quantity of the pollen in successive layers at Lunkaransar, the wettest period was from 3000 to 1800 BC. At the end of that time, just as the Harrapan civilisation crumbled, Lunkaransar lake turned salty and began to dry. The radiocarbon dates may be subject to correction, but the coincidence is not. We can leave it to the archaeologists to wrangle about the coincidence, and the relative importance of rivers and local rain in the Harrapan economy. The gist of the discovery is that the summer rains visited this part of India for many centuries, during which time the inhabitants prospered; then the monsoon broke its promise, allowed the trees to perish, and gave the land back to the sand dunes that possess it now.

For human beings today, the monsoon rains are the most important trick of the weather machine. They support the largest populations, in China and India. If you

A mass of cold air sweeps across an airbase in Florida from the side of a thunderstorm. It forms an arc cloud which is usually harmless – unless it collides with another storm and sets the air spinning in a tornado.

add the other countries of south and east Asia that are utterly dependent on them, and take into account the lesser monsoons of Africa which supply the Nile as well as the countries of East Africa and West Africa where the rain falls, half mankind relies on the monsoons' gift of water. And, each year, farmers from West Africa to China wait anxiously for the rains to come to their thirsty lands. The weather machine always manages the great overturn that delivers the rain, but sometimes imperfectly. Even in the densely populated Yangtse valley of China the rainfall in July can be twice as great in one year as in another; at the desert fringes the variation is much greater.

In central India a dry wind blows from China for six months of the year; then for six months it comes from the ocean, bearing moisture. The onset of the monsoon rains is the most important date in the farmer's calendar – about 12 June in central India. By then, rain has been falling for two weeks in the south of the country but, although the air is already humid over central India, the rain hesitates. At such times, in the past, villagers might yoke their maidens to the plough, to bring on rain; or suspect that the grain merchant had set his daughter to work with a spinning-wheel of dead men's bones, to unwind the clouds and put up the prices. In 1972, the monsoon was three weeks late and India lost almost a third of its food production; 1973 was a good year; in 1974 the rains were again late in some parts of India, while Bangladesh suffered fatal floods.

The summer monsoon is a trade wind that overshoots the equator. The huge land mass of Asia distorts the pattern of the global winds, and this is the only sector of the weather machine where there is a big exchange of air between the northern and southern hemisphere. In winter, the air pumped to the top of the tropical clouds of the Indian Ocean flows almost exclusively northwards; it returns at the surface as the winter monsoon, cold and dry off the land. The mountains of Tibet and the Himalayas, though, shield India from the cold of Siberia, which in winter is colder than the North Pole. But in summer, Asia begins to rival the equator as a warm zone. Although Tibet and the Himalayas seem cool to us, they are in fact a good deal warmer in summer than the air normally is 15,000 feet or so above sea level. The warmed air rises there, more vigorously than at the equator, and flows towards the southern hemisphere, passing high above India and the ocean. The return flow comes from the southern hemisphere as the trade winds; but they make a right turn as they cross the equator (another effect of the Earth's rotation) and sweep over India from the south-west.

It is like a sea-breeze that blows across the beaches on to the warm land on a hot summer afternoon, but on a gigantic scale. Asia in summer is hotter than the equatorial Indian Ocean – that is the basic point. Off East Africa cold water wells up from deep in the ocean and that reinforces the temperature difference between land and sea. At present a great deal of interest surrounds the question of whether the patterns of sea temperatures between Africa and India may govern, in some way, the quality of the monsoon and explain its variations from year to year. Opinions differ widely, though, and there has been a great lack of information about sea temperatures.

The Geophysical Fluid Dynamics Laboratory in Princeton is the US government's prime centre for using computers for research into weather and climate. There, an outstanding Japanese theorist, Syukuro Manabe, has been able to reproduce the main features of the monsoon by calculation. The next chapter will look more closely at computer 'models' of the Earth's climate; for the moment we need only note that Manabe can do experiments to see what contribution various geographical features make to the monsoons. For example, if he abolishes the Himalayas, the rains do not penetrate so far to the north. An experiment done by a colleague from India, Jagadish Shukla, shows the importance of the mountains of Burma in walling in the moist air and making the rain spill back, as it were, on eastern India and Bangladesh. And Manabe and Shukla find, in their model monsoon, a striking dependence on the temperature of the sea surface off the east coast of Africa. Other theorists of the Indian monsoon have suspected such an effect. The computer suggests that, when the sea is relatively cool, less moisture finds its way across the equator to sustain the monsoon rain.

A supply of moist air from across the Indian Ocean is not enough to produce rain. The air over the desert in Rajasthan is moist enough, but the rain does not fall. It has to be released by the air rising, so that the water vapour condenses and forms rain. In Rajasthan, as in other deserts, the air has a persistent tendency to sink. Elsewhere in India the rain is released by mountains forcing the air upwards, or by depressions, travelling regions of rising air.

Two or three depressions a month track slightly north of west across the country from the Bay of Bengal. At any one time a monsoon depression can be causing very heavy rains over a tenth of all India. Much effort by Indian meteorologists goes into forecasting these monsoon depressions and into research aimed at understanding their tracks, because they determine the actual rainfall. A worrying problem is the tendency, especially in August, for the monsoon depressions to veer northwards away from the densely populated valley of the Ganges, to drench the Himalayas instead. That causes a break in the monsoon rains which, if it persists, can wither the crops.

The monsoons are very much a part of the system of global winds. In winter a branch of the jet stream, divided by Tibet, flows from the west across northern India; in summer the upper winds flow from the east. The easterly jet guides the monsoon depressions across India, while wriggles in the jet help to determine where they form. The pressing human need to anticipate, if possible, whether the monsoon rains each year are going to be adequate or not, whether India and other countries will or will not avoid famine, has led to all sorts of attempts to relate the outcome of the monsoon to the weather in distant parts of the world. Repeated failures suggest that local conditions, in the Indian Ocean and Tibet, for example, may be the factors that determine the quality of the monsoon. The monsoon is more of a governor than a servant of the world's weather. For example, the fact that July and August are the wettest summer months in Britain may be connected with the summer switch of the westerly jet from India to north of Tibet, which seems to shift the zigzagging jet stream south over Europe.

More general arguments about changes of climate, which do not purport to predict the performance of the monsoon in a given year, may still be valid. These predict weaker monsoons if the climate in the northern lands becomes colder. One reason may be that the 'false equator' of the land mass in summer has less advantage in temperature over the equatorial ocean. Another possibility is that cooling in the north and less rainfall in the monsoon lands may both be symptoms of a general weakening of the global winds.

Be that as it may, there are grounds for optimism in the fact that India and China have supported great civilisations for many thousands of years, while the climate elsewhere has fluctuated widely. The desert of Rajasthan is fortunately unusual enough to be a tourist attraction. But the pessimists have a case, too. They point out that intermittent famines have long been a grim part of the Indian and Chinese way of life, killing untold millions of people. The populations of those

As the wind sweeps from the south-west off the Indian Ocean, the monsoon clouds bring life-giving rain to southern Asia – and the ever-watchful satellite records them.

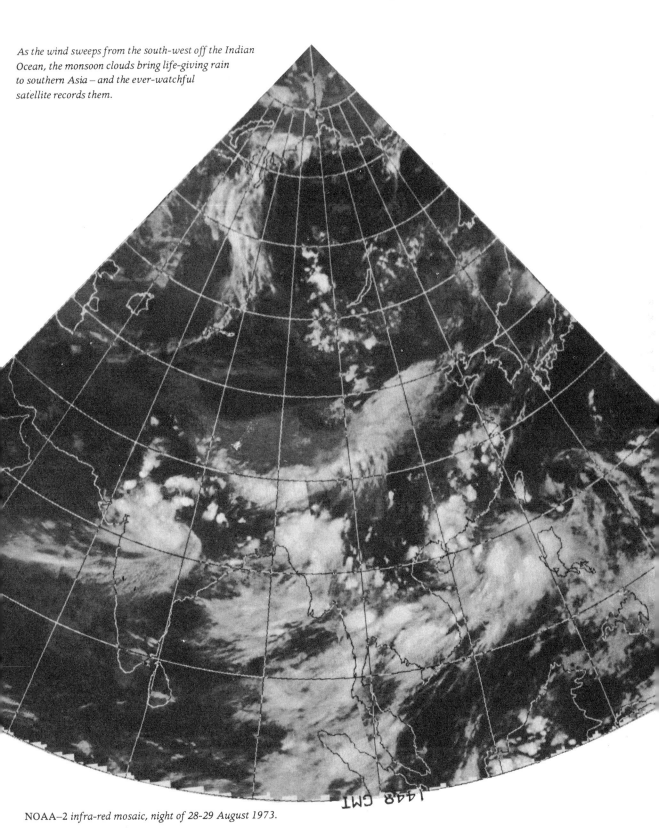

1448 CMT

NOAA–2 infra-red mosaic, night of 28-29 August 1973.

The summer rain is usually a matter for rejoicing in India – unless it comes in floods, or too little or too late.

countries have grown to their present huge numbers during the untypical climatic period of the 20th century, when the monsoon rains were unusually reliable.

In 1974, the Indian Meteorological Service was denying any trend in rainfall in India to match the warming in the north up to 1950, and the subsequent cooling. That conclusion was based on total annual rainfall at many stations. But if you look at the information another way, the story is very different. Every year in India the pattern of the summer rains varies, with some districts suffering drought and others having plenty of rain or too much. The disastrous years are those when drought occurs in almost all regions of India. From the Indians' own records, such disastrous years occurred fairly often at the beginning of this century, became very rare in mid-century, and have since become commoner again. By 12-year periods, 1901-72, there were three bad years in the first period (1907, 1911, 1912), three in the second (1913, 1918, 1920), two in the third (1928, 1929), two in the fourth (1939, 1941), only one in the fifth (1951) but again three bad years in the sixth period (1965, 1966, 1972). Moreover, the timing of the monsoon, which is crucial for planting and harvesting, is less reliable than it was at mid-century. A succession of droughts in India could be the most deadly event since the Second World War.

A turn for the worse

The vultures have been feeding well along the southern fringe of the Sahara. In 1973 several countries of the Sahel of West Africa reported the death from drought of 30 to 70 per cent of their cattle, as well as widespread crop failures. *Sahel* is an Arabic word meaning shore, used here for the inland southern 'shore' of the Sahara desert, a zone across Africa where human subsistence is just about possible. Closest to the deep Sahara range

the nomadic herdsmen, but farmers, too, scratch a living from a few months' meagre rainfall. For them the present change of climate is not a statistical curiosity or intellectual speculation but a toll of dead animals and the sight of their soil blowing away in the wind. Ten million people are refugees from the drought.

The rains of the Sahel come in from the south in summer, as wedges of moist oceanic air insert themselves over Africa in miniature monsoons. The rainfall is abundant along the coast from Sierra Leone to Nigeria, and in southern Ethiopia, but it peters out towards the north in the zone where farming and herding are marginally possible. Reliable and timely rain comes less far into Africa than it did in recent decades. The marginal zone has therefore been shifting towards the sea, with the Sahara advancing southwards behind it. Rainfall in the Sahel has been 'below average' for seven or eight years – that is how the reports are worded. But here, as in many other parts of the world, good meteorological records exist only for the past few decades, when the climate has certainly not been 'normal'. Even in this century, there were severe droughts in the Sahel in 1900-3, 1911-14 and 1930-1. In the better period after 1931, which coincided with the decades of maximum warmth in Europe and the Arctic, the herdsmen edged further north with their cattle, towards the heart of the Sahara. The farmers advanced behind them and populations grew.

Now the Saharan tide has turned. The herdsmen are driven back on to farmland, whether or not the farmers have yet abandoned it. The towns are crowded with hungry refugees. The traditional strategy of the Tuareg nomad in hard times is to turn to trading to recoup his losses, but amid general impoverishment conditions that is hard to implement. Nevertheless, the nomads are more adaptable than the farmers, and can readily move north or south according to the season's rainfall.

It may be that only they can hold on to the threatened lands of the Sahel.

The tragedy has been starkest in northern Ethiopia, in the eastward extension of the Sahel zone. The disaster exposed the feudal abominations of the 'hidden empire', which hid its starvation for as long as possible and neglected to feed the dying from its own resources. Not until October 1973, when more than 100,000 were already in their graves and many more were past help, did the Ethiopian government admit that 3-4000 were perishing every week in the northern districts and appeal for help from other countries. Pneumonia or gastroenteritis or cholera was often the immediate cause of death, but hunger was at the root of it.

West Africa faced the drought more openly and the relief supplies were at least more timely; even so, many thousands died. Since March 1973 the countries most affected have coordinated their efforts to cope with the disaster. Although human starvation seems to have been kept within bounds so far, millions have suffered great deprivation, and disease has taken an even tighter grip than usual in that part of Africa. Besides the cholera, typhoid, diphtheria and measles, one sign of desperation has been the migration from the central plateau of Mali into the notorious valley of the Black Volta, where river blindness is rife.

Four countries lie wholly within the affected zone – Mali, Upper Volta, Niger and Chad. They could become uninhabitable if the desert were to continue its southward march during future decades. We may be witnessing the elimination of these nation states by drought, while nearby Senegal and Mauretania could be reduced, for practical purposes, to coastal strips.

Meanwhile, human pressures on the land do nothing to impede creation of new desert. Herds of cattle may overgraze the land and loosen the soil, though not so readily as do the cultivators, who create dust bowls when the rainfall is inadequate. Cutting firewood is certainly harmful near the towns of the Sahel. Yet environmentalists are probably adding insult to injury when they blame the local people for *causing* the southward expansion of the Sahara. On the northern edge of the Sahara, in the Atlas mountains of Algeria and Morocco, rainfall has been increasing. That suggests that the whole dry zone of the Sahara is moving bodily southwards and that the Sahelian drought is a consequence of world-wide changes of climate.

The simplest and most ominous theory is that the advance of the Sahara in the Sahel is directly linked with the general cooling of the northern hemisphere that has been in progress since 1950. As the Arctic zone of cold enlarges, the stormy zone and desert zone shift southwards towards the equator. It is ominous because the cooling in the north shows little sign of relenting. By the record of past changes in climate, such phases can last for many decades or even centuries. According to one forecast, the desert will have advanced another 40 miles southwards by the end of the century. The forecaster declares that, in the face of so strong a natural trend, to try to stem its progress by massive irrigation schemes would be futile. The people of the Sahel will simply have to leave and try to make a living in the wetter coastal states of West Africa, such as Ghana.

A more optimistic view has it that the global connections do not work in this fashion, as far as the Sahel is concerned. The drought may then represent an unfortunate 'run' of bad years, perhaps associated with peculiar weather over the Mediterranean, but not a part of a long-term global trend. In that case the drought could end at any time (1974 was wetter) and it need not recur for several decades. And there are more positive opinions about the possibility of rolling back the desert. By increasing the vegetation on the desert margin, one might actually promote increased rainfall in the area.

Senegal, 1973, and the dead cattle of the Sahel. Will the drought persist, or may it be already ending?

The lives and well-being of many people are at stake in the Sahel. This case illustrates all of our climatic anxieties: the uncertainty about how greatly human activity does or could affect the climate; the urgent need for reliable forecasts of climatic changes; and the hazards of inaccurate or speculative forecasts. The only remedy is to discover the root causes of climatic change all over the world. The next chapter will not tell you what they are, but it describes the present attempts to find them.

A storm over Nigeria. How the weather in different parts of the world interacts is a matter of urgent international research.

2 Causes and Effects

The fashionable place to be during the summer of 1974, if you were a weatherman, was in the tropical Atlantic off West Africa. An armada of 40 research ships from 10 nations, including 14 Russian ships, was gathered for the biggest single experiment in the history of science. At the airport at Dakar research aircraft laden with instruments shared the runway with occasional aircraft still carrying relief supplies to the drought-stricken lands to the east. Dakar is the capital of Senegal, itself one of the countries most affected by the southward advance of the Sahara Desert. For a hundred days in 1974, Dakar was the headquarters for GATE – the Atlantic Tropical Experiment of the Global Atmospheric Research Programme. The intensive scientific work was relevant to the problem of famine in Africa only in a roundabout way, but it signalled a new phase in human efforts to understand the weather of the tropics.

GATE brought together many of the strands of the new meteorology to knit a net of instruments and brainpower around the clouds of the tropical Atlantic. Most conspicuous, perhaps, was the international character of the enterprise: apart from the muster of ships and aircraft, weather stations in 57 countries from Arabia to Peru were making essential observations of conditions in the upper air by meteorological balloons. Then there was the use of weather satellites, with the Americans launching, just in time for the experiment, the splendid new GOES satellite. It remained poised over the area; besides showing everyone what the clouds were doing, its endless stream of pictures enabled researchers at the University of Wisconsin to deduce the winds blowing in the tropics from the movements of the clouds. The main props of the new meteorology are the big computers, and nine ultra-powerful machines in the USA, Canada, Britain and Japan, as well as smaller ones in other countries, had been made ready to digest and interpret the vast quantities of information flowing from the GATE area and to offer forecasts that could be checked as the weather unfolded. Not the least significant aspect of GATE was the importance attached, however belatedly, by the weathermen of the rich northern countries to the processes in the tropics.

Their effort matched their interest. Take just one of the GATE armada, the Canadian weather ship *Quadra*. She was pulled off her station in the Pacific for the occasion. Her powerful radar was wanted for finding the rainstorms and summoning aircraft to investigate them. In addition she carried more than 60 sensors trained on sea and sky, as well as instrumented balloons for releasing into the upper air and an instrumented 'fish' that swam 600 feet deep behind the ship, recording conditions in the sea. *Quadra* was one of the group of ships deployed in and around the main area of the experiment – a box of the ocean 800 miles out from Dakar.

The chief aim of GATE was to 'catch' in this box a succession of the great clusters of clouds that occur in the tropics. The tropical oceans are in fact the main boiler houses of the whole weather machine, absorbing the Sun's rays and heating the air from below. The cloud clusters are strung out like rows of smoking chimneys across the oceans.

The pumps of the tropics

With hindsight you could say that the cloud clusters had been staring meteorologists in the face ever since the first satellite pictures of weather over the tropics became available in 1960. Tropical weather was thought to be incurably random, with the wind blowing where it pleased and clouds popping up like bubbles of steam in a pot of boiling water. Sailors had reported long days of thundery rain and fresh winds. Weathermen in the tropics were well aware of the recurrences of very big rainstorms and some of them had concluded, without the help of satellites, that their clouds were organised in clusters. But the general pattern was unknown.

Committees are never supposed to come up with anything new, least of all a scientific discovery. The rule was broken in October 1968, when a group of research meteorologists met at the University of Wisconsin. They were to study satellite pictures of tropical disturbances for the whole of 1967, as part of the preparation for the forthcoming Global Atmospheric Research Programme. One of India's liveliest meteorologists, P. Rama Pisharoty from Ahmedabad, was the convenor; the other members of the group were Ted Fujita from Chicago and Michio Yanai from Tokyo, but Verner Suomi (the host) and a dozen others from American laboratories lent a hand. Their stocktaking of the behaviour of cloud clusters is enshrined in an appendix to a GARP planning document.

Satellite pictures clearly show that clouds over the tropical oceans form clusters, each several hundred miles across, with wider areas of clear skies in between. On a typical day there are about 30 cloud clusters scattered around the Earth's tropical zone. Over the Pacific, the Atlantic and the southern Indian ocean, cloud clusters are typically 100 to 600 miles across. The most densely populated clusters are 250 miles across. They all travel westwards at an average speed of 12 knots. Many of them form at night and each survives, as a rule, for two to seven days. Over the Indian Ocean and India during the monsoon rains, huge cloud clusters occur up to 1200 miles across. They also form over the mountains of East Africa. Some travel westwards, just like the oceanic cloud clusters, and when they reach the Atlantic they may weaken or strengthen. But some African ones travel eastwards and others stand still. Over South America, cloud clusters are rare;

ordinary local thunderstorms and mountain-made clouds are the rule.

From below, each cluster is a widespread area of rain and thunderstorms. The thousands of individual clouds comprising it have shorter lives than the cluster as a whole. A small minority of them are very large thunderclouds that do most of the work. That work is to take the warm, moist air of the tropics and pump it upwards through 40,000 feet or more. Cooling as it rises, the air forms first water droplets and then ice crystals, and a big cloud makes plenty of rain. As the cloud strains the moisture out of the air, it releases heat which helps to drive the column of rising air even higher. By the time it reaches the top of a big cloud the air is very cold and very dry, in contrast with the warm, muggy air at the surface. It has also expanded into the very thin air of the higher atmosphere. But that air still has plenty of energy by virtue of its great height and rarefaction – potential energy, a little like that of water at the top of a waterfall. When it eventually comes down, the air is compressed again and that makes it hotter; thus the energy is recovered.

Indeed, so vigorous is the pumping in the cloud clusters that some of the lifted air has to sink again right away, in and around the cluster – otherwise there would be a vacuum. The hot dry air returned to the surface becomes a sponge to absorb yet more moisture from the warm ocean. But much of the lifted air travels north or south for 2000 miles and comes down in the dry zone at the edge of the tropics. With the same hot, sponge-like qualities it then flows back towards the equator in the trade winds, sucking moisture out of the sea as it goes. Eventually it encounters a new cloud cluster forming, which it can feed afresh.

So the overall action of the cloud clusters is this: from the oceans warmed by the tropical sunshine and fanned by the trade winds, the air extracts warmth and

Follow the tropical edge of this satellite view of the northern hemisphere and you will see the 'necklace' of cloud clusters, of different shapes and sizes, right around the Earth. They inject enormous energy into the weather machine, and some turn into tropical cyclones.

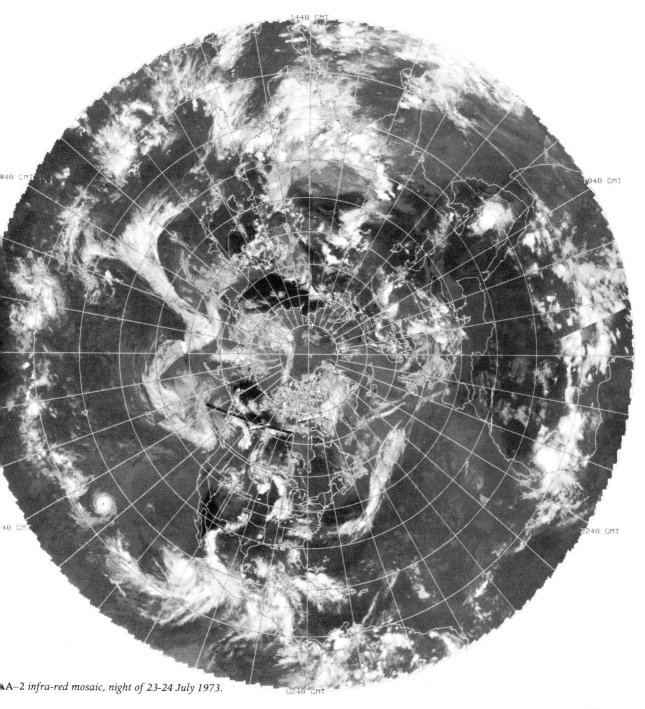

▲A–2 infra-red mosaic, night of 23-24 July 1973.

A British research aircraft that took part in GATE. Its nose boom puts the measuring instruments into undisturbed air.

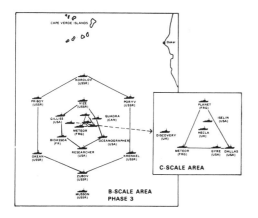

GATE 1974 – the Atlantic Tropical Experiment of the Global Atmospheric Research Program – involved ships, aircraft, satellites and weather stations deployed across a wide belt of the tropics ('A' scale, far left). Research ships of the 'B' and 'C' scales, off West Africa, set a tighter trap for cloud clusters.

moisture and the cloud clusters convert them into rain at the surface and a huge weight of lifted air high above it. Weathermen have known for a very long time that something of the sort had to go on in the tropics. But the organisation of this activity in the cloud clusters, plainly visible in the satellite pictures, means that the meteorologists can see, and monitor, the process at work all around the globe. They find that the process is not a continuous one, which it could be if the clouds were scattered at random. Instead it is occurring, at any one time, over only relatively small parts of the tropical oceans. Both locally and all around the tropical world, the vigour of the cloud clusters varies from day to day and week to week. This machinery that injects energy from the tropical oceans into the world's air pulsates, like a great piston engine.

In short, the system of tropical cloud clusters is the most significant piece of the weather machine to be discovered since the jet streams. When the world's meteorologists were planning to collaborate in finding out more about the tropics, the cloud clusters were plainly the prime target. The area first picked for the experiment was the Marshall Islands, for the very good reason that the Pacific Ocean in their vicinity has the highest productivity of cloud clusters anywhere in the world. Political and logistic difficulties led to the experiment being transposed to the Atlantic, as GATE.

The hope was that generalisations might be drawn about the behaviour of the cloud clusters and their constituent clouds. The mathematical descriptions of the weather machine, whether for forecasting by computer or for analysing the climate, cannot take account of each cloud individually or even of the smaller cloud clusters. Rules are needed to estimate the flows of energy, moisture and whirliness from cloud clusters, and to predict when and where and how frequently the cloud clusters will appear. Plainly that should help in

making better forecasts for the tropics themselves. Many people living in the tropics depend on the rain that the cloud clusters bring and often suffer disastrous floods or droughts. And some cloud clusters turn into hurricanes or typhoons.

As the cloud clusters deliver their pulses of energy to the global winds, they contribute to the weather all over the world. Indeed that was supposed to be the chief justification for so much effort in GATE. Better knowledge of the tropical cloud clusters would enable the northern countries that supplied the ships and aircraft, satellites and computers, to forecast their own weather for several more days ahead. At the time of writing that hope is still being tested, as all the information gathered during GATE undergoes interpretation. One of the crucial issues, in fact, in current efforts to understand the world's weather and climate, is: which parts of the machine are in charge?

Does a great cloud cluster spring up because all that energy in the warm tropical oceans and the moist air above them is bursting to escape, willy-nilly? Or is the tropical zone working as a coherent system, tied together by waves travelling westward around the world, much as the zigzags in the main jet stream impose their pattern on the weather of the stormy zone? Regularities in the cloud clusters suggest that to be at least part of the story.

Beyond the organisation of tropical clouds into clusters, the clusters in turn form patterns. They tend to lie along bands running across the oceans parallel to the equator, with gaps between successive clusters about twice as wide as the clusters themselves. Although the cloud clusters pump energy into the atmosphere their operations seem to be governed by the condition of the atmosphere as a whole.

That is why, while the ships in the main experimental area in GATE were cruising under the storms and the

research aircraft were weaving among the thunder-clouds, weathermen in a great belt reaching nearly half way around the tropics, across Africa, the Atlantic and South America, were reaching out with their balloons into the upper air. The search was for variations in the winds and pressures. Among these might be identified zigzags or waves progressing around the tropics and creating the upwards suction that allows cloud clusters to form. The mountains of East Africa and the equatorial Andes may play a part, like the Rockies in the main jet stream, in helping to generate the pattern. So far, so good; but the awkward question is still to come.

Is the pattern of cloud clusters in the tropics subject to influences from north and south? In other words, perhaps the stormy zones or even the polar zones beyond are really in charge of what happens in the tropics, as if they might switch the tropical clouds on and off to satisfy their own needs for energy. Weathermen are divided into two camps on this issue, which is far from being settled, although what governs what is going to be crucial to the diagnosis and forecasting of changes in climate.

Meteorologists contradict one another almost as much as psychologists do. Both professions confront systems of exquisite complexity, the weather machine and the human brain. Their disagreements are not a sign of stupidity. They are due in part to lack of knowledge, but even more to genuine contradictions within the systems themselves. A man's mood can affect his hormones, while his hormones can influence his mood. Where do you break into the circle to see why he is unreasonably angry or depressed? The weather machine is replete with similar feedback links. To take just one of the most obvious and most intractable: the temperature at a place helps to determine how cloudy it is, but cloudiness in turn has a big effect on temperature.

This chapter deals with the somewhat desperate search for causes and effects in the weather machine. If the weather did not vary from year to year you could dismiss the problem. You could say that the machine worked steadily as a unit, like a cruising car. Even a simple car engine has feedback. The sparks, for instance, that explode the fuel and turn the engine are themselves generated by the engine. They are nicely synchronised with the valves and no further explanation is needed for why the car keeps going. The weather machine, unfortunately, is like a car that keeps changing speed and there must be reasons. What are the internal controls (like the throttle and gears) or the external controls (like road conditions and hills) that account for the changes?

Meteorology grows up

The conversation of the weathermen of the 1970s would make only limited sense to their predecessors of half a century ago. The assurance with which the experts today speak of conditions 'aloft' and of weather on the other side of the world would amaze the old school. Casual references to 18-level numerical models or polar-orbiting satellites would lose them completely. Fifty years ago, when weather forecasting was still much more of an art than a science, a few far-sighted men knew that the weather would have to be calculated – that only arithmetic could interpret the poetry of turbulence that we call the weather. They lacked the machines for doing it.

The long pre-scientific phase of weather-watching was strong on poetry but weak in understanding. For the Greek mythmakers, the winds were lusty gods, including Boreas the north wind who in winter seemed to fertilise the Athenian countryside in preparation for the spring. In China the Taoist pantheon had a bureaucratic flavour: the woodcutter Hsin Hsing, for ex-

ample, was immortalised as assistant secretary of state in the Ministry of Thunder. Amid this charming dross were nuggets of good sense. Anaximander, 26 centuries ago, knew that rain represented the return of water evaporated from the Earth; the older *Rigveda* of India had pronounced that 'out of the sun comes rain'. The way mountains assist in rainmaking was also well known in ancient times. But understandable confusion left doubts about where the weather ended and outer space began, and the Moon and the stars were widely supposed to take part in the weather.

In 1643, after the death of his teacher Galileo, Evangelista Torricelli invented the barometer and meteorology became vaguely scientific. His instrument showed that the invisible air weighs heavily on the Earth's surface – one ton on every square foot. It revealed that the weight can vary from day to day, in accordance with the weather, as masses of air shift their ground. Thermometers of the modern type for measuring the temperature of the air also made their appearance in the mid-17th century, in time to register the record chills of the next 200 years. And human hair, which shrinks when it is damp, provided the means of measuring the humidity, or moisture-content, of the air.

The shortcomings of meteorology in the *era of the weather glass* are enshrined in every home that possesses a barometer of the traditional kind. The inscriptions such as 'fair' alongside the high-pressure end of the scale and 'stormy' at the low-pressure end are approximate to say the least. The chief value of the solitary 'glass' was for sailors. A rapid change of pressure up or down (although tradition emphasised the 'falling glass') is a sign of strong winds. But for understanding the processes that make and change the weather, the lone meteorologist is helpless. Only when observations made simultaneously at different places are compared do the relationships between pressures, temperatures

and wind become plain. The first weather map was drawn in 1820 – but it was based on observations collected from 39 observatories by the Meteorological Society in Mannheim 37 years earlier! From this and other maps meteorologists discovered the patches of high and low pressure called anticyclones and depressions, and the rotary nature of the winds around a big storm. But when the information had to be gathered by mail these observations at scattered places were of no value for weather forecasts.

Not until the *era of the telegraph* could modern weather-forecasting begin. Within a few years of its invention in 1843 the telegraph was beginning to carry routine weather observations. During the ensuing decades meteorologists began to gain some instinct for the way patterns shifted on their charts from day to day, and forecasts steadily improved. Paradoxically, when the Norwegians were cut off by the First World War and had to rely on ingenuity rather than the telegraph, they made the discovery that the characteristic storms of Europe and North America are regions where warm tropical air and cold polar air collide. Another important discovery, from pioneering balloon flights in France early in the 20th century, was the stratosphere – a layer warmer than expected that starts 25,000 to 59,000 feet up and puts a lid on the clouds and the weather. But the patterns that marched across the weather maps remained almost as mysterious to meteorologists as the clouds that moved across the sky were to the ancients.

Then came the *era of the radiosonde*. Starting in the 1930s, but with enormous acceleration to meet the needs of military flying during the Second World War, routine flights of expendable balloons disclosed the condition of the air high above the ground. The balloons carried packets of instruments – miniature weather stations – which transmitted their measurements to

'Nature's weather map.' In this view from space, one can make out storms on both east and west of North America, the dry south-western states, and tropical cloud clusters at the bottom. The eastern Atlantic is in shadow. (Apollo-10, May 1969.)

the weathermen on the ground. When balloons could be followed by telescope they also gave information about the winds at high altitudes; nowadays they carry radar reflectors and can be tracked even through cloud. These upper-air observations, and the weather maps for different heights that they made possible, at last enabled the weathermen to make sense of the surface winds, by connecting them with the winds aloft.

All that was preamble. Meteorology of today has its roots in the developments of the 1960s as much as in anything that went before. We are in the *era of the computer* – but also in the era of the satellite and of unprecedented international cooperation. In the World Weather Watch organised by the World Meteorological Organization, virtually every country on Earth takes part and for the simplest of reasons: the weather is a global process and everyone needs everyone else's information. In the war against bad weather the quarrelsome human species is more truly united than in any other activity.

Twice a day, at noon and midnight Greenwich Mean Time, designated stations all over the world launch their radiosonde balloons. There are about 700 such 'upper-air' stations by land and sea in the Global Observing System. In addition 9000 other meteorological stations, 6000 merchant ships and dozens of aircraft send in their routine weather reports. Despite strategic gaps, especially in the southern oceans, the effort is impressive – not to mention the half-million balloons literally thrown away every year, and often mistaken for flying saucers.

And twice a day, by the 'hot lines' of the Global Telecommunications System, great tides of weather information flow around the world. In principle, at least, Moscow knows what is happening in Tahiti within three hours. At twenty-one Regional Meteorological Centres and three World Meteorological Centres (Washington, Moscow and Melbourne) the incoming information is collated and forecast charts for huge areas are generated. These are then available to any national weather service that wants them. The meteorologically advanced countries are happy to swap the products of their computers for the weather reports on which the computers feed.

Although the computer is the characteristic tool of the times, it was the satellite that burst more suddenly upon the meteorological scene. The first weather satellite, *Tiros 1*, went into orbit in 1960 and sent back pictures of the Earth's cloud cover that astounded even the meteorologists. They expected chaos; instead they saw, as one of them described it to me, 'nature's weather map, drawn for us'. Depressions and tropical storms stood out clearly. Even today skilled observers are still making discoveries about the workings of the weather machine just by looking closely at satellite pictures. One recent example is the way patches of clear sky surrounded by clouds in the morning are liable to be the scene of thunderstorms later in the day.

For an investment of a few thousand dollars anyone, anywhere, can set up a receiving system and pick up the automatic picture transmissions (APT) from American weather satellites. The cloud pictures are not the best that the satellites obtain – those go to special ground stations on command – but many countries make use of the APT pictures for instantaneous impressions of the weather around them. From high-quality pictures, wind speeds can sometimes be deduced by comparing the positions of clouds in successive pictures. By special treatment of the images, to increase the contrast, meteorologists can identify very deep clouds that are probably raining. In addition, the satellites show snow lying on the ground, and ice on land and sea around the Arctic and Antarctica.

Most weather satellites fly on polar orbits – that is to

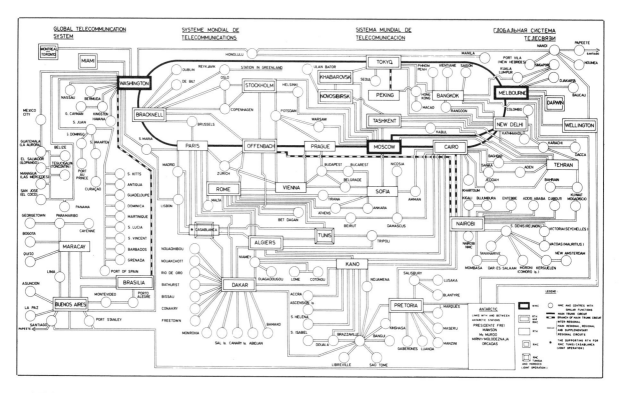

'Hot lines' of the World Weather Watch (above). The Global Telecommunications System pools the observations from all parts of the world; it also distributes the analyses and forecasts made by the big computers. The dark line denotes the trunk circuit linking Washington, Moscow and Melbourne.

Above right: in Hong Kong, a miniature radio telescope is trained on an American weather satellite passing invisibly overhead. Anyone can pick up pictures of the weather of his area, as seen from space, just by investing in the necessary ground equipment.

Left: A weather balloon goes up during hail-prevention operations in Georgia, USSR.

say around the Earth over the poles, so that as the Earth rotates the whole of it is gradually scanned on successive orbits. Of the American satellites, those called successively *Tiros 1* to *Tiros 10* and *Nimbus 1* to *6* have been for experimental purposes, testing new instruments and furthering basic meteorological research. Starting in 1966, *ESSA 1* to *9* and *NOAA 1* to *4* have been operational systems for the weather forecaster. The Soviet operational weather satellites are called *Meteor*.

Besides these satellites orbiting over the poles, so-called 'geosynchronous' weather satellites have begun to play a part. They are satellites orbiting essentially around the equator at a height of 22,000 miles; at that distance the satellite takes exactly 24 hours to orbit the Earth, so that it keeps in step with the Earth's rotation and therefore always remains poised over the same region. From its vantage point it can observe about one quarter of the Earth at once. The first American *ATS* satellite of this type was launched in 1966 and a year later, from above the American tropics, *ATS 3* began

returning spectacular colour television pictures every 24 minutes, until the colour system failed. The new *GOES* satellite launched in 1974 has already been mentioned, in connection with GATE.

The modern satellites open up the possibility of a new kind of weather service – not forecasting but 'nowcasting'. General weather forecasts made by computer cannot predict individual clouds and showers, yet small-scale weather in the form of winter snowstorms, summer hailstorms or tornadoes, can be extremely serious for the communities in their paths. High-quality satellite pictures can show these features, and track them across the countryside.

Verner Suomi of the University of Wisconsin has long been a leading pioneer of weather satellites and his research group has been developing the technologies of computers and visual displays for 'nowcasting'. As each new satellite picture comes in, showing important clouds forming or moving, the aim is to produce, within minutes, useful information about winds and weather for instant dissemination to local weather stations and,

by television, to the public. Suomi's group is experimenting with various ingenious displays of maps, pictures and written information. These include a system like the 'instant replay' used for sports events, to show successions of satellite pictures for the district; then the viewer can see for himself whether a storm is heading his way.

Satellite pictures by visible light depend, of course, on daylight, but some satellites, including *GOES*, take pictures also by heat rays (infra-red). These work better at night, when the Sun interferes less. The ground, the sea and the clouds all emit heat rays even in darkness. From the intensity of the radiation their temperatures can be assessed. Although the pictures returned to Earth by the satellites are still immensely valuable, other instruments are supplying numbers that can be fed into computers.

Very accurate sensors are now being carried by satellites that measure the heat rays or very short radio waves emitted from different levels of the atmosphere. In principle they provide a means of seeing the temperature of the air at all levels, all over the world. There are difficulties in the way the heat-ray emissions from different levels confuse one another, in the effects of clouds, and in interference from gases, particles and water vapour even in clear air. But what the satellites lack in accuracy at present they make up for in their global coverage. One day satellites may replace the expensive network of stations and ships which observe the upper air by balloons. They also promise worldwide measurements of the surface temperature of the sea, which could not be obtained in any other way.

Weather by numbers

Twice a day, in some of the world's most powerful computers, model tempests rage across model oceans, while model jet streams zigzag over the mountain-tops of the model Earth. They exist as figments only, as transient numbers in the feverish electronic arithmetic of machines racing to make their model weather develop more rapidly than the real weather. If they did not run faster than the atmosphere they model, there would be no forecast. Although the computers have to cut corners to stay ahead in the race, they are already a little better at making weather forecasts than are the most experienced human beings. In theory the computers could do very much better still.

No mystery surrounds the ability of a computer to represent the atmosphere in a numerical model. The weather is predictable only because the world's winds, for all their seeming unruliness, conform to physical laws. The computer programs embody many of those laws – the effects of the Earth's rotation, for example, the way the air must move in response to varying pressures and temperatures, and the release of energy when moisture condenses to make clouds. The machine is also instructed in the rudiments of the Earth's geography. But even though the programs by necessity make many approximations and simplifications, the calculations are still prodigious. That is why the weathermen always demand the most powerful computers available and why computer forecasts have become really practicable only in the past few years, as the commercially available machines have caught up with the weather problem.

The following brief description of numerical forecasting is based on current practice at Britain's Meteorological Office at Bracknell. It is representative of the state of the forecasting art in the best equipped operational centres.

During the three hours after noon or midnight Greenwich Mean Time, hundreds of thousands of pieces of weather information flow into the terminals at

To generate a weather forecast for Britain, the computer reckons the weather right across the northern hemisphere, calculating events at every one of the 3000 crosses. It then proceeds to more detailed forecasts for Europe, using the finer, dotted grid.

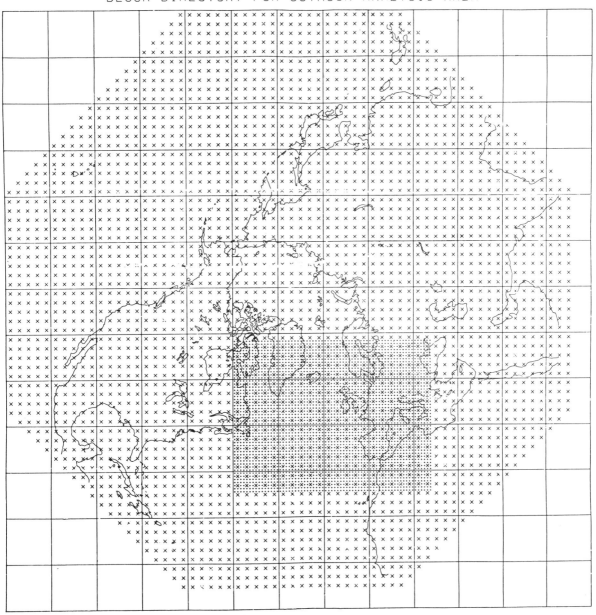

BLOCK DIRECTORY FOR OCTAGON ANALYSIS AREA

Bracknell from all around the world. From this information the computer deduces the present state of the atmosphere throughout the northern hemisphere, at ten different levels above the ground. Sometimes a satellite picture gives more accurate information about the position of a particular feature – a storm centre, say – and there is special equipment for telling the computer about it, by correcting a visual display of the computer's analysis of the present weather.

The computer then uses its programs to calculate the future changes in the weather at three thousand points across the northern hemisphere. After 20 minutes its prediction for 24 hours ahead is ready. The machine continues and produces, after another 20 minutes, a forecast for three days ahead; by then it has done 10,000 million sums. Finally it carries out a more detailed computation for western Europe, with the aim of predicting rainfall in particular.

Meanwhile the responsible humans have been at work with the 24-hour forecast. The computer's predictions for the upper air normally go straight to the airlines, untouched by human hand. But the surface forecast demands more thought. The human forecaster does not tamper with the computer's forecast unless he has good reason to do so – but quite often he does. He is aware of certain kinds of errors that the computer programs can fall into; he also knows about the quirks of the nation's weather and about important features, like thunderstorms, that may be beneath the computer's notice. So there is give and take between the computer's capacity for arithmetic and the human forecaster's judgment. He will study the movements of the weather systems shown by the computer, perhaps adjusting them a little and adding details. In any case the rather abstract general forecast has then to be translated into simple terms that everyone can understand, for region by region around the country, and that, too,

is largely a matter for human experience.

The central objective in meteorology today is to improve the computer products, both by making the forecast for tomorrow's weather more reliable and by enabling the machines to forecast successfully for a week or more ahead. These two objectives are closely linked. According to André Robert, who leads Canada's efforts in numerical forecasting, the present sources of error can be pinpointed. For a three-day forecast for a region in the middle of a continent, as much as 45 per cent of present error is due to the computing procedures. Contemporary computers have to take short cuts in the calculations, in order to produce a forecast quickly enough. But the speed of computers is still increasing so rapidly that Robert expects this source of error to be virtually eliminated by 1980. For forecasts of more than five days, effects spilling over from the southern hemisphere become important, so Robert expects the forecasting calculations to be done for the whole globe rather than just the hemisphere, as in present practice.

The second source of error is lack of accurate data on the present weather around the world. Robert estimates the damage it does at 20 per cent of the total error. A very big increase in the number of weather stations, far beyond anything envisaged at present, would be needed to reduce the error significantly by conventional means. Robert is hopeful about the perfection of satellite instruments, which can cover the whole globe.

The remainder of the error, a substantial 35 per cent, comes from a source that is not easily stoppered. It is insufficient knowledge of the physical processes that make the weather: the uptake of moisture from the sea, the effects on the winds of hills and the texture of the landscape, the conversion of a cloud into rain, and many other effects. The numerical forecaster has to be able to specify the calculations necessary to take account of such processes. They have to be done in a general way, because reducing everything to detail could take far too much computing time, but they must still be accurate. The belief that better knowledge of the atmosphere would lead to better forecasts inspired the present lavish international efforts in meteorology.

They had their origin in a single sentence in a speech by President John Kennedy to the General Assembly of the United Nations in September 1961: 'We shall propose further cooperative efforts between all nations in weather prediction and eventually in weather control.' A rough blueprint for a World Weather Watch and an international research programme had already been prepared by a group of American meteorologists. Within a year an American and a Russian working together (Harry Wexler and V. A. Bugaev) had 'internationalised' the proposals. The UN General Assembly called on the World Meteorological Organization and the International Council of Scientific Unions to develop the observational and research programmes.

The Americans continued to set the pace, in particular with a report by the National Academy of Sciences in 1966, on the feasibility of a 'global observation and analysis experiment'. Some weathermen still refer to the multinational enterprise as 'Charney's experiment', a reference to Jule Charney of the Massachusetts Institute of Technology. As a pioneer of computer models of the weather he played a dominant part in formulating the ideas. During the next few years the Global Atmospheric Research Programme (GARP) began to take shape, and many countries joined in the preparations, and began to adapt their national research plans for meteorology to suit the global ambitions.

The Japanese had the first big experiment under way by February 1974, in and around the Ryukyu Islands, with the headquarters in Okinawa. It was called the Air-Mass Transformation Experiment, and it observed the 'transformation' of a bitterly cold 'air-mass' that blows off China in winter and meets the warm sea of the Kurishio current flowing northwards from the tropical Pacific. As the cold, dry air passes over water 20 degrees warmer than itself, it sucks up enormous supplies of warmth and moisture and becomes stormy. The same sort of encounter occurs off the eastern United States and in the Mediterranean, so the experiment had world-wide significance. Off China in winter the process is exceptionally vigorous. Arrays of instruments set up on the islands, on ships stationed out at sea, and in aircraft that flew into the clouds, were deployed to sense the fingers of rising air carrying away the energy of the sea and then dispersing with the wind or concentrating into storms.

Next came GATE, looking closely at the cloud clusters of the tropical Atlantic in the summer of 1974, as the international programme gathered momentum. Impressive though that effort was, it was just one more step towards a bigger objective. So great is the eventual commitment of men, ships, computers and money to the Global Atmospheric Research Programme that even some of the participants tend to forget that it is just an experiment, albeit on a huge scale. Already some fairly firm promises of computed weather forecasts for one or two weeks ahead have been made to encourage governments to supply the necessary funds. In fact the overall purpose of GARP is to see whether the weather is really so predictable.

In a speech as president of the Royal Meteorological Society in London in 1967, George Robinson urged his colleagues to remember the possibility of failure. Even large-scale features of the weather may be inherently unpredictable beyond about five days. For smaller-scale features of the weather, such as thunderstorms, the limit of prediction is much less – a few hours – and Robinson assailed an assumption of the planners that there was a clear separation of the 'large-scale' and 'small-scale' weather. He argued that there was no gap and that today's large-scale weather features simply dissipate themselves in small-scale motions. They lose their identity and become no longer predictable. He estimated the time for this to happen at five days – the practical limit of reliable forecasts in areas already well supplied with observations before the big experiment began. Robinson concluded his speech: 'I hope it will be shown that I have made a mistake.'

To say that the atmosphere obeys strict physical laws does not guarantee predictability. Another sceptic is Edward Lorenz, a colleague of Jule Charney at the Massachusetts Institute of Technology and himself a distinguished theorist. He emphasises a peculiar characteristic of the weather and its mathematical descriptions, that an effect can be as disproportionate to its cause as a forest fire started by a cigar butt. In weather forecasting it means that a very slight difference in the prevailing weather – the presence or absence of a thunderstorm for example – could make a substantial difference to the world's weather a few days hence. Because large-scale motions of the atmosphere are fairly stable in their behaviour, discrepancies do not matter very much over a period of a few days. But a thunderstorm can double its intensity in 20 minutes and any discrepancy on that scale can grow just as rapidly. In practice thunderstorms are too small to register in the network of global stations, or to figure in the computer's calculations. But they typify the subliminal sources of error that can eventually frustrate attempts at long-range numerical weather forecasts.

Small errors in today's observations of the state of the world's weather can be equivalent to overlooking a storm. Their effects, too, can become greatly magnified as the forecasting calculations proceed. Considerations of this kind have led Cecil Leith of the US National Center for Atmospheric Research to try the so-called Monte Carlo method, in experimental forecasts. As its name implies, this means making a succession of trials with random variations in them. Leith varies the starting conditions slightly and compares the different forecasts that the computer produces. They give an impression of the range of possible outcomes, allowing for the uncertainties. In practice they would help the forecaster to hedge his bets.

Atlantic spells

However erratically and unpredictably the great weather systems may develop and move, sooner or later they meet our general expectations of sunshine and rain. Indeed, half the problem of explaining why the climate changes is to say why it does not change very much. Somewhere in the machine there are masters that discipline the weather, and those masters are likely candidates to be the agents of climatic change.

An approach to long-range weather predictions which is completely different from numerical forecasting sets out to identify some of the factors that determine the general behaviour of the weather machine. I mentioned earlier the apparent effect of the sea temperature off East Africa on the quality of the monsoon of India. The unusual spells of good or bad weather that all of us have experienced at some time or another can be due to peculiarities in the temperature of the ocean many thousands of miles away. The sea temperature is thus one of the masters of the weather. But it is also a slave, because the weather affects the sea temperature.

Jerome Namias was a pioneer of 'extended forecasts' for the US weather service; today he pursues his interest in sea temperatures at the Scripps Institution of Oceanography in San Diego. He has to work with very sketchy information. It comes largely from reports by merchant ships working the usual routes: they measure the temperature of the water taken in to cool their engines. If satellites fulfil their promise of being able to measure the sea-surface temperature anywhere in the world, life will be easier for Namias and his followers. Meanwhile, despite the difficulties, he has been able to spot patches of unusually warm or cool water in the world's oceans and to see far-reaching connections with weather around the world. A difference of only a degree or two in the water temperature can have very striking consequences.

In the summer of 1972, according to Namias, abnormal patches of warm and cold water in the North Atlantic were responsible for much grief. The winter of 1971-2 had been unusually mild in the eastern United States and as a result the Atlantic water east of the United States was exceptionally warm. At the same time the water off Greenland was exceptionally cold; this was associated with an extraordinary crop of icebergs, the greatest number in the North Atlantic since 1912 when an iceberg sank the liner *Titanic*.

These patches of water then affected the jet stream. The contrast between them greatly strengthened the jet stream and also diverted it far to the north. That promoted and sustained a strong zigzag in the jet stream. It guided warm, dry weather over much of the North Atlantic and also – one zigzag further around the world – over western Russia. There the 1972 grain harvest failed because of the drought – helping to push up world food prices. But other legs of the same zigzag pattern brought cold, wet weather to the eastern United States and the British Isles, both of which had dismal summers in 1972. To crown all, in June the low-pressure area over the eastern United States sucked in Hurricane Agnes. That storm, which caused disastrous flooding, took more than 100 lives and did damage estimated at more than $3000 million – the most costly storm in American history. The sequence of events in

1972 is only one, though perhaps the most dramatic, of the interactions between sea temperatures and the weather investigated by Namias.

For the long-range weather forecasters in Britain, the temperature of the sea off Newfoundland is the best single indicator of the weather to come in the weeks ahead. A group at the Meteorological Office is charged with publishing, twice a month, a weather forecast for a month ahead. They have searched through the meteorological records, classifying every month of every year since 1888 according to eight possible oddities of the Atlantic sea temperatures a month before. So, knowing the present sea temperatures, they can see what the precedents say for the month ahead. Warm water south of Newfoundland in September, for example, puts low pressure over Scandinavia in October; in the same period cold water off Newfoundland puts high pressure over the British Isles. The basic reason is that the water off Newfoundland is the birthplace of most of Europe's depressions, and warm water encourages them. In winter, when the sea temperature is least changeable, independent assessors judge this group's forecasts to be at least moderately good nine times out of ten; in summer, when the sea has less control, the score is two out of three.

Far more effort goes into making the long-range forecast than just checking the Atlantic sea temperatures. Although these are rated the most important indicators there are many others. The sea temperatures of the Pacific may be reinforcing or conflicting with the effects of the Atlantic water. Sea temperatures in the coastal seas can affect the risks of fog or frost. The extent of sea-ice and snow is very influential; for instance, the date of the onset of spring in Britain is quite well related to the amount of ice between Greenland and Iceland. The general state of the global winds and the jet stream and its zigzags, and any tendencies the

The disastrous northern summer of 1972. The contrast between very cold water off Greenland and warm water further south tended to lock the jet stream to the route shown. It swung far to the north of the Soviet grainlands, leaving them with a devastating drought. Meanwhile, its southward swerve over eastern North America conspired with Hurricane Agnes to cause floods. The underlying 'map' is a specially prepared sum of the satellite pictures for 90 days that summer. To make it easier to read, it is here printed as a negative so that grey areas are cloudy and white areas are dry.

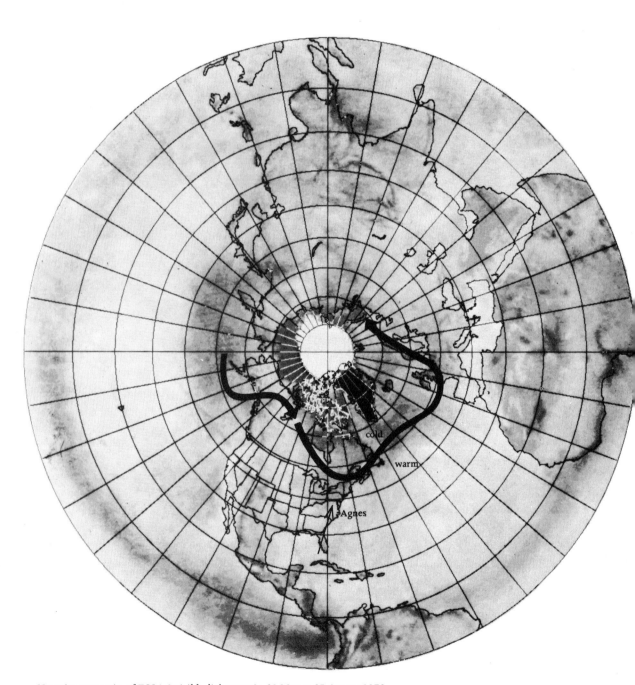

Negative composite of ESSA-9 visible-light mosaic, 28 May to 25 August 1972.

weather may have to become set in a particular pattern, are carefully considered. And the forecasters make painstaking comparisons with past years, looking for weather situations that are analogous to the present, by all sorts of meteorological tests. The same group at Bracknell makes experimental forecasts for three months ahead, which are not yet released to the general public, although I understand that they have a success rate comparable with the monthly forecasts.

The sea, like the air, is a swirling fluid. It has its own 'weather' and sometimes its own calamities. In 1971 Peru had the largest fish-catch in the world, and the anchovies teeming in the waters off western South America were much in demand as animal feeding stuff. By April 1973, the Peruvian government had closed the fisheries. What had happened in the meantime was the visitation of El Niño, the Christ Child, so called because it usually occurs around Christmas; in 1972 it struck in January. In normal times, a strong current flowing from the south eases away from the coast near Peru. In the gap fertile water wells up from the depths of the ocean. If the current shifts inshore, it blocks off the fertile water and the young anchovies may fail to thrive, although the ecological consequences are complicated. In 1972 the catch fell from the normal 10 million tons to 4·5 million tons, mostly of older fish.

Some 6000 miles away in the Pacific Ocean, a great current of eastbound water flows just north of the equator against the prevailing wind. It carries warm water towards South America, at up to 40 million tons a second. But sometimes it almost stops. The simplest way to detect its erratic behaviour is to measure the level of the sea on various tropical islands. When the current flows strongly, the sea-level falls at islands north of the current and rises at islands to the south of it – a matter of a few inches. The Earth's rotation makes the water tilt a little as it moves. Klaus Wyrtki of the

University of Hawaii has found that El Niño off Peru coincides with surges in the current in the mid-Pacific; the same warm water affects the weather in central America and the USA.

Some meteorologists argue that El Niño happens because the trade winds weaken, and that the mid-Pacific current, which runs against the wind, goes faster for the same reason. As researchers try to see how the pieces of the weather machine mesh together around the world, issues of cause and effect become sharper.

Around the equator, where the cloud clusters pour out their moisture on the rain forests of the Amazon and the Congo, there are deserts or near-deserts. For example, air that rises over Indonesia sinks over the parched Galapagos islands on the other side of the Pacific, off the coast of South America, where the water is cold except during El Niño. Changes in the positions of rising air and sinking air around the equator, or intensification of the movements, are responsible for unexpected droughts and floods. Drought afflicted Brazil during El Niño of 1972; by 1974 the pattern had changed and Brazil suffered disastrous floods at the end of March, in which thousands perished. Associated with these shifts are changes in the ocean currents but also changes in weather and water patterns in the stormy zone. Which is the cause and which the effect among so many, widely scattered parts of the machine?

The grand old Norwegian of meteorology, Jacob Bjerknes, pioneered the frontal theory of depressions half a century ago. He spends his retirement in California pursuing the changes from year to year in the huge expanse of the equatorial Pacific, between South America and the international dateline. He sees feedback at work between the winds and the ocean water. Warm water promotes cloudiness and updraught, drawing in stronger trade winds, but the winds help to bring cold water to the top which cools the ocean surface. The

Rain around the tropics. The equatorial air has exchanges of its own around the globe, rising in some places (wet) and sinking in others (dry). Slight east-west shifts in this pattern can greatly alter the rainfall at particular places. (After H. Flöhn.)

The ocean surface hides great flows of underlying water, as this simplified section through the Atlantic shows.

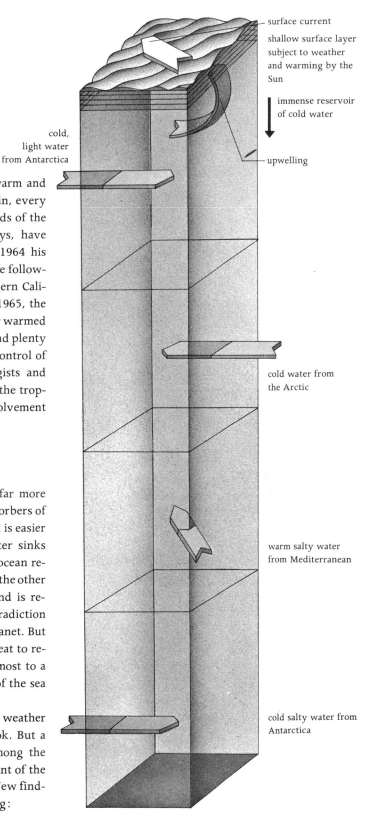

surface current

shallow surface layer subject to weather and warming by the Sun

immense reservoir of cold water

upwelling

cold, light water from Antarctica

cold water from the Arctic

warm salty water from Mediterranean

cold salty water from Antarctica

result is, he says, a rhythmic change from warm and wet to cool and dry conditions and back again, every two years, which is plain in the rainfall records of the Pacific islands. These changes, Bjerknes says, have world-wide consequences. For example, in 1964 his Pacific equatorial band cooled down and in the following winter, with weaker global winds, Southern California and Spain, for example, were dry. In 1965, the Pacific winds having also slackened, the water warmed up again, and Southern California and Spain had plenty of rain. Such is the 'equatorial' view of the control of the weather machine, but other meteorologists and oceanographers look for the controls outside the tropics. That question remains open, but the involvement of the oceans is not in doubt.

Oceans, ice and deserts

The oceans are far more massive and store far more energy than the air, and they are the chief absorbers of the Sun's rays. Warming the surface of the sea is easier than cooling it, because chilled surface water sinks and is replaced by water from below. So the ocean resists cooling at the surface. The atmosphere on the other hand resists heating, because hot air rises and is replaced by cooler air from above. This contradiction may provide one of the thermostats for our planet. But once the sea is cool, it needs a great deal of heat to rewarm it; indeed the weather would come almost to a halt in summer, were it not for the memory of the sea (and ice) for the cold of winters past.

The oceans are so important a part of the weather machine that they would merit half this book. But a great overturn of ideas now in progress among the oceanographers would make an orderly account of the ocean currents either obsolete or premature. New findings on at least four basic points need digesting:

1. The ocean currents may play a much larger part than anyone had suspected, in transferring heat from the tropics towards the poles.

2. The ocean currents may be mainly driven, not by the winds as has been widely supposed, but by differences in temperature and saltiness from place to place.

3. The ocean water forms layers, no more than inches thick, each distinguished from the layers above and below by sharp changes in temperature and saltiness; they must greatly affect the up-and-down movements of the water.

4. The oceans contain a great number of powerful eddies, equivalent to the travelling depressions of the air, but much smaller (only 60 miles or so in diameter) and persisting for months on end. These feed the major currents, just as the airy eddies contribute to the jet stream.

Discoveries about the eddies in the early 1970s mean that many more observations may be needed to know the 'weather' of the ocean reliably. From the point of view of weather and climate the oceans are turning out to be appallingly difficult to deal with. The temperature of the sea surface can persist for months and then change in the course of a single storm; while the deep water takes a thousand years to undergo any appreciable change. The circulating currents of the ocean surface suck up water towards the surface from below –

but there are three miles of water to draw on. The deepest water of the ocean is naturally the heaviest. At present it is the very cold, salty water from the edges of the Antarctic ice. If there were no ice, the warm and very salty water flowing out of the Mediterranean, the Red Sea and other places of very high evaporation would cover the ocean bottom.

To explore the interactions of the atmosphere and the upper layers of the ocean elaborate experiments are planned by the Americans for the Pacific (NORPAX) and by the British for the Atlantic (JASIN). As for the oceans themselves, POLYMODE in the Atlantic is to look further into relationships of the eddies and the main ocean currents, while the International Southern Ocean Study should dispel some of the remaining ignorance about the waters around Antarctica.

The influence exerted on the weather by ice on the sea, and by snow and ice on land, is no less marked than the effects of unusual sea temperatures. Ice and snow reflect the sunshine rather than absorbing it; the result is an abrupt fall in air temperature compared with nearby air over ice-free sea or land. This temperature contrast tends to guide the jet stream and its attendant storms parallel to the edge of the ice. The behaviour of ice is yet another Pandora's box of complex problems. Ice sheets and glaciers flow very slowly downhill and eventually melt. But they persist for hundreds or even millions of years and change only slowly. Meanwhile

The deep Sahara. According to a new theory, the brightness of the barren desert itself encourages the air above it to sink. That prevents the formation of clouds and rain, so that the desert in a sense makes itself.

The maps below show a big switch in the winter weather over North America, brought about by a change in the pattern of sea-surface temperatures which affected the route of the jet stream. (Dark shading, cold water; light shading, warm water.)

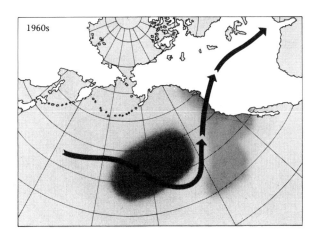

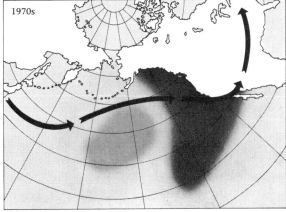

they have profound effects: chilling the air, bottling up water on land and so lowering the sea-level, and despatching great icebergs into the ocean as ambassadors of coldness.

Pack ice can pass heat from the sea to the air only by slow conduction through the ice, but there are often gaps or 'leads' in between the ice floes. The sea ice can thicken and spread, or melt and contract; it reduces the effect of wind on the sea; and since ice contains less salt than the sea water does, when it forms or melts it alters the composition of the sea surface. Besides melting, snow and ice also evaporate. The sunshine in July in the central Arctic is sufficient to melt away a 20-inch thickness of ice.

The area of sea ice in the Arctic decreases during the summer by about a quarter from its maximum extent at the end of winter. In the Antarctic the variations are far greater: in winter the sea ice covers an area seven times larger than in summer, and from being much smaller than the Arctic sea ice it becomes half as large again. During winter open ocean near one of the poles keeps the region warmer than it otherwise would be, but curiously the presence of ice in summer reduces the loss of heat by the atmosphere – by keeping the air cold and thereby reducing its emission of heat rays.

The Sahara, Arabia and the other scorching deserts act like ice sheets as far as the atmosphere is concerned. They are light in colour and reflect back into space most

of the Sun's energy falling on them. As a result the air high up over the deserts receives less heat than its entitlement. In 1974, Jule Charney looked afresh at the conventional idea that the sinking of the air over the deserts, which keeps them dry, is due to the rolling-over circulation from the equator, which makes the trade winds (see page 22). He calculated that the cooling effect of the desert brightness produced five times more sinking than the trade-wind circulation would cause, unaided. If the desert carried vegetation it would absorb much more heat and push back the sinking air. In that sense, Charney told me, 'the desert causes itself'.

The air is heated from below, by the warmth of the Sun absorbed by the surface of the land or sea. The pattern of heating, over warm and cold parts of the ocean, over the sub-tropical deserts, the high mountains and the polar ice, is the biggest factor in determining the overall pattern of the world's climate. Changes in the pattern of heating can alter the normal climate, for a week or for ten thousand years. That is why 'thermal forcing' is now the watchword of those trying to understand fluctuations of climate.

If a block causes persistent weather that alters the Earth's surface, its consequences can be felt long after the jet stream has reasserted itself and abolished the block. A block over Greenland in winter increases the production of sea ice around Iceland and the chilling effects of that ice will still be felt in the following spring; the same is true of heavy snowfalls in the Rockies brought about by a block over Alaska. Wind and weather also help to warm or cool the sea, which in turn can affect the air long after the cause of its own warming or cooling has been removed, and can even help new blocks to form.

Some weathermen think that, with so many interactions to play with, the climate can change of its own accord, without needing any outside influence. Like an economic system going through booms and slumps, the weather machine can judder. For a start, odd-numbered years tend to have one kind of weather, even-numbered years another. In other words there is a two-year rhythm. It is most conspicuous in the stratosphere over the equator, where the wind blows east one year and west the next. In the lower atmosphere there is a two-year rhythm in the strength of the main jet stream, for example, and in such diverse symptoms of weather as the floods of the Nile and the quality of European wine. Bjerknes's alternations in the water temperature of the equatorial Pacific could be involved.

Jerome Namias thinks that the climate goes into a 'régime' for perhaps ten years and then switches abruptly to another régime. Taking the winters in the eastern United States, he sees two switches, one 'down' and one 'up', since the early 1950s, when the winters were warm. In 1957 a great revolution occurred in the sea temperatures of the North Pacific. For example, a warm patch north of Hawaii was replaced by a cold patch, and warmer water than before appeared off the west coast of the United States. The winter of 1957-8 was cold in the eastern United States, and that régime, in the Pacific sea temperatures and the cold eastern winters, persisted for fourteen years. The exceptionally cold winter of 1962-3 in the eastern United States and western Europe Namias attributes to an extensive warm patch north-west of Hawaii. In 1971 another big change

occurred in the sea temperatures of the Pacific, since when the eastern United States and western Europe have enjoyed warmer winters again; the western United States is cooler.

Namias is one of those investigating changes of climate who strongly question the idea of a systematic trend, including the cooling trend in the northern hemisphere which others are so confident in reporting. The east-west shifts of weather patterns, associated with the altering zigzags of the jet stream, do indeed complicate the problems of identifying and following global, as opposed to local, changes of climate. If the pattern happens to put warm air on land where there are plenty of weather stations and cold air on the oceans or wastelands where there are few (or vice versa) the observed trends in temperature can be quite misleading about global warming or cooling.

Subjectively, too, the warm winters that many of my readers will have experienced in the 1970s may make them understandably sceptical, as Namias is, about any continuous change towards a cooler climate in the northern lands. For the would-be optimist, Namias's thesis about alternating régimes, which redistribute heat rather than losing it, is one of the stronger straws at which to clutch. But the argument can go the other way: benign meanders of the jet stream may be temporarily concealing from us the real trend. The fact remains that indicators which owe nothing to the deployment of weather stations speak of continuing cooling – most conspicuously the increase in snow cover and sea ice in the Arctic.

A leading American polar scientist, J. O. Fletcher, argues that the climatic changes in the north, such as the Little Ice Age, the warming in this century, and the present cooling, may be produced by contrary changes around Antarctica. Icier conditions in Antarctica seem to be associated with a strengthening of the global

'The human volcano.' According to some experts, man-made smoke and dust is helping to cool the Earth at present.

winds; on the other side of the equator, strong global winds help to warm the northern lands. In 1920-50 Antarctic explorers had to contend with worse pack ice and intense winter cold, even while the ice was receding around Greenland and Iceland. Considerably more snow fell at the South Pole. Since then the situation has reversed, with the Antarctic growing warmer even as the Arctic has cooled.

Although the weather machine may well be capable of self-engineering the longer periods of warming and cooling, there is no virtue in ignoring outside influences that may have helped. These we turn to now.

Outside influences

Mt Agung, the big volcano of the Indonesian 'paradise' island of Bali, erupted violently in March 1963. Fine dust and sulphurous vapours, flung high into the atmosphere, were caught up by the high-level global winds and carried around the world. Magnificent sunsets reddened the tropical sky, because of the multitude of minute particles, typically 50 millionths of an inch in size, lodged in the stratosphere. The dust warmed the high atmosphere; at 70,000 feet above the tropical Atlantic it was three degrees warmer than usual. The volcanic dust was spread by stratospheric winds towards the poles and, by intercepting the sunlight, it cooled the whole world. In 1964 and 1965 temperatures at the Earth's surface were apparently reduced by an average of one-third of a degree.

Mt Agung's was the biggest single volcanic eruption of this century; Hubert Lamb estimates its 'dust-veil' at 80 per cent of Krakatao's after the notorious explosion of 1883. A big eruption affects the world's weather for a few years after the event, until the dust settles. Apart from the direct effects of cooling, it probably weakens the global winds. The fatal summer in Scotland in 1695 followed four substantial eruptions in Indonesia in 1693-4. In the 100 years from 1750 to 1849, the culmination of the Little Ice Age, Lamb estimates no fewer than ten dust veils greater than Krakatao's, including three enormous ones from Mayon in the Philippines (1766), Tambora in Indonesia (1815) and Coseguina in Nicaragua (1835). Of these, Tambora can be said to be the origin of *Frankenstein*, because it was the atrociously wet summer in Switzerland in 1816 that provoked Lord Byron to suggest that Mary Shelley should write a story. Her gloomy thoughts matched the weather.

Spates of volcanic dust are sometimes very much a fact in climate, but their inherent irregularity makes them an unlikely cause of persistent trends in the climate, such as the transitions into and out of the Little Ice Age. And the present cooling in the north has begun in the mid-20th century, a period remarkably lacking in big eruptions.

The possibility that the Sun itself flickers, so that its energy supplies to Earth's atmosphere vary from year to year or from one century to the next, has to be taken seriously. Direct evidence is almost non-existent. Prolonged observations by satellite, which might testify clearly to changes or constancy in sunshine, have not been undertaken.

Because the weather on the Sun might well affect the weather on Earth, the conversation among climate experts always comes back, enthusiastically or rudely, to sunspots. The dark blemishes on the bright face of the Sun have been known to astronomers for centuries. A sunspot is a relatively cool area created by an enormous storm like nothing known on Earth – a great magnetic vortex in the hot, electrified gas of the Sun. Every eleven years or so the number of spots on the Sun rises to a maximum and then falls away again.

Despite 300 years of enthusiastic searching, the

Dust from volcanoes sometimes affects the weather world-wide.

symptoms of 11-year cycles in the Earth's weather remain surprisingly scanty. The air pressure may be slightly higher over the Arctic when sunspots are abundant. At times of maximum storminess on the Sun, the great explosions called solar flares are more frequent and there are hints that they can cause existing storms in the Arctic to become a little more intense.

The sunspot theory of weather itself goes through cycles of approval or disrepute among weathermen. At present it is fashionable in the Soviet Union, where at least one eminent meteorologist lost his job in 1973 because he did not have sufficient faith in sunspots. In Britain the long-range forecasters use the sunspot cycle with great caution as just one of very many possible indicators; they look at the weather of ten or twelve years ago for hints as to how this year's weather might develop. (They make 10-year or 12-year comparisons rather than 11-year ones because they take the two-year cycle into account as well.)

The Sun also goes through a magnetic cycle; its magnetism reverses its direction during each sunspot cycle, which means that it takes two sunspot cycles – 22 years – to return to the starting direction. On Earth, rhythms of 22 years' duration are just about apparent in the rate of growth of trees in Labrador and in the strength of the global winds. Droughts in the Great Plains of North America have recurred every 22-23 years, but they do not match the double sunspot cycle, which itself varies in duration.

More significant for those who invoke the Sun to explain changes in climate over the centuries are slow changes in the storminess of the Sun. These are indicated by changes in the number of sunspots at successive peaks of the sunspot cycle. There are corresponding changes in the precise duration of the sunspot cycle, which can vary from 10 to 14 years and is shorter in stormier periods. These climatic changes in the Sun go

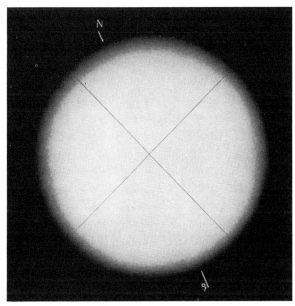

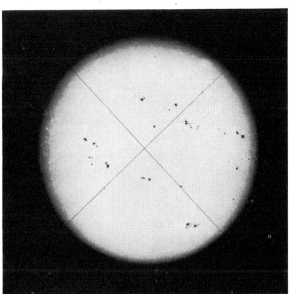

Solar weather alternates between clear and spotty conditions.

through cycles but there is more than one rhythm. One is of about 80 years' duration, with recent peaks of storminess in the late 18th century, the mid-19th century and the early 20th century. Another is a rhythm of about 200 years. It is hard to pin down in the systematic records of sunspots, which go back only to 1700. It becomes plainer, in a curious way, in discrepancies in the dates of archaeological remains. And by an involution in our story we are led back to tree rings.

Radiocarbon dating depends on measuring the amount of radioactive carbon in the remains of living things. The carbon dioxide that plants absorb from the air as they grow contains traces of radioactive carbon. After a plant dies the radioactive atoms decompose one by one. The less the surviving radiocarbon, the older can the remains be judged to be. But all sorts of errors began to appear, most alarmingly in the radiocarbon dates for Egyptian sites for which reliable historical dates were available. Plainly the radiocarbon in the air had not been constant. So the method had to be turned the other way around to correct the radiocarbon timescales, by measurements in wood that could be dated precisely using its tree rings. In the 1960s the fluctuations were exposed.

The amount of radiocarbon in the air has varied from century to century by a few per cent. It was a bonus for the climatologists as well as the archaeologists. The radioactive carbon is made by cosmic rays, atomic bullets that come from the Sun and from distant exploding stars. Their arrival on Earth is governed by changes in the activity of magnetism of the Sun. A 200-year rhythm goes back to Roman times, with recent peaks in radiocarbon occurring around 1300, 1500 and 1700; before 300 BC, the rhythm looks more like a 400-year cycle. Intense production of radiocarbon corresponds to cool conditions on Earth.

More striking than these cycles, because you do not

have to peer to find it, is a particularly big increase in radiocarbon production during the Little Ice Age, compared with the warm period that preceded it. The peak came in AD 1500. What is more, peaks in radiocarbon production also occurred at the times of other great advances of the glaciers, around 1000 BC and 3500 BC, in keeping with the 2500-year glacier cycle of George Denton (see page 12). Here is one of the best pieces of evidence, albeit indirect, that climatic change in the Sun may indeed be responsible for changes in the climate on Earth.

The Earth's magnetism may help to fix the route of the zigzagging jet stream – that is another recent offering about climatic change. The strongest magnetism in the northern hemisphere occurs close to the magnetic pole in the Canadian Archipelago, but a second region of strong magnetism lies over northern Siberia. And these are two regions to which the jet stream habitually gives a wide berth, travelling far to the south of them. A theory first advanced in Germany in 1952 and newly revived says that this is more than a coincidence. In the 17th century, in the depths of the Little Ice Age, the magnetic pole was due north of London. That means (by this theory) that the jet stream was thereby pushed to the south in the vicinity of Europe, exposing Europe to cold Arctic air. Another suggestion is that the mean temperature at a given place is closely related to the prevailing strength of the Earth's magnetism at that point, and that when the magnetic field grows stronger the temperature falls, and vice versa.

Pet theories, together with the natural vagueness in the rhythms (even if they are real), make climatic change one of the untidiest areas of modern science. As knowledge and ignorance stand at present, the most plausible cause of climatic change in recent centuries is the changing Sun, with interwoven rhythms of about 80 years, about 200 years and about 2500 years.

Here it is worth recalling the rhythms of about 80 and 180 years detected in the record of the Greenland ice by Willi Dansgaard's laboratory. On top of these primary changes, presumed to be Sun-given, are cool periods produced by unpredictable volcanic eruptions. The judder in the weather machine itself allows the jet stream, the oceans and the polar ice to play games with the Earth's environment. And now, as the joker, comes 20th-century man's ability to rival the processes of nature, for good or ill.

The human volcano

Man-made dust, according to Reid Bryson, is a sufficient explanation of the present changes of climate, from the Arctic cooling to the African drought. Bryson is a meteorologist and climate specialist who works at the University of Wisconsin's Institute of Environmental Studies. He is happy to be a controversial figure among both the weathermen and the environmentalists, and his views make a convenient starting point for reviewing human effects on the weather.

Bryson adopts the theory that natural volcanoes have been the prime cause of the changes of climate in past centuries, but says that human influences have become marked. The strong warming during the early part of this century is explained by Bryson as due both to a waning in volcanic activity and an increase in the carbon dioxide in the air by man's burning of coal and oil. As scientists have recognised for a long time, carbon dioxide probably warms the Earth, by the 'greenhouse effect'. Like glass, it lets sunlight pass to the Earth's surface, but absorbs some of the heat rays, at infra-red wavelengths, emitted by the warm Earth. Water vapour acts in a similar way.

After 1930, the 'human volcano' supervened, according to Bryson. Growing numbers of busy people all over the world were producing smoke and dust, not just from industry but from careless agriculture too as the wind caught up the dust from overgrazed or over-worked land. This addition to the air's natural burden of dust has, according to Bryson, overcome the warming due to carbon dioxide. As a substitute for natural volcanoes, it has produced both the cooling in the north and the droughts of Africa and India. The prognosis is famine and the fault is all mankind's. Meteorologists dispute much of Bryson's detailed reasoning about how carbon dioxide and dust work in the atmosphere. His grim prognosis might be right even if the reasoning were faulty, and his thesis of cause and effect – the human volcano – is not easy either to refute or to confirm satisfactorily.

In a forest south-east of Stockholm, in 1974, ecologists and meteorologists were measuring the liveliness of the trees. Towering over the forest was a mast 150 feet high carrying, at various levels, instruments for wind, temperature, humidity – and carbon dioxide. For Bert Bolin of Stockholm University's Institute of Meteorology the carbon dioxide measurements were of global significance. As a leader of international efforts to get to grips with the problems of climate, Bolin recognises the possibly crucial importance of man-made carbon dioxide and the greenhouse effect. It is a tricky business, though.

Natural processes remove carbon dioxide to make limestone on the ocean floor, while the ocean water itself contains fifty times as much carbon dioxide as the air does. The increase in carbon dioxide in the air during the last century has been ten per cent, but that is only about half of what you would expect from the amount of coal and oil burned. The rest has presumably been absorbed in the oceans – or else incorporated into trees. Increased carbon dioxide promotes plant growth.

During each day, the instruments on the mast in the Swedish forest record a decrease in the carbon dioxide in the air just above the treetops, as the trees absorb the gas and combine it with water to make the chemical materials they need for growth. At night they register an increase as the trees respire, 'breathing out' carbon dioxide much as animals do. Meanwhile, the forest represents a vast amount of carbon dioxide that, for the time being, has been taken out of the air. When the trees die and rot, or are made into paper and eventually burnt, carbon dioxide returns to the air. This give-and-take between the air and living things is part of the intricate process which determines the final balance of carbon dioxide in the air and hence the strength of the greenhouse effect.

Even the supposed warming effect of the carbon-dioxide greenhouse cannot be taken for granted. If any warming resulted in increased cloudiness, intercepting the sunlight, that could hold the warming in check. An increase in cloudiness by two per cent could offset an increase of 100 per cent in the carbon dioxide in the air. As Bolin sees it, the effects of carbon dioxide will eventually be evaluated in elaborate numerical models that show the interactions of many different factors affecting the climate, of which carbon dioxide and cloudiness are only two.

At the Mauna Loa Observatory in Hawaii you can watch the whole world breathing, using detectors that record the carbon dioxide in the air. Because there are more plants in the northern hemisphere than in the southern, the carbon dioxide diminishes during the northern summer as the plants grow, and increases again during their resting phase in winter. This observatory, run by the US National Oceanic and Atmospheric Administration, is one of the strangest laboratories I know. As if at a moonbase, dozens of automatic instruments, breathalysers for the planet, sprawl on the chunky lava. The air is thin, 11,000 feet high on the

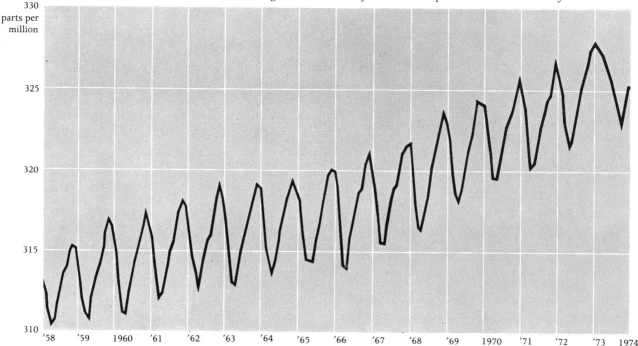

Down the tube at the Mauna Loa Observatory, Hawaii (right), samples of clean air come for measurement of the carbon dioxide in it. The record (below) shows 'the whole world breathing' with an annual rhythm, but also a persistent increase over the years.

flank of the Mauna Loa volcano, but it is one of the cleanest places on Earth; upwind, the nearest land is 2000 miles away. So here you can check on the lasting, truly global consequences of pollution. Similar stations are being set up at the South Pole, in Alaska and various other places, for the world-wide monitoring of human effects on the environment. Carbon monoxide, sulphur gases, lead – whatever it is, Mauna Loa has a detector. One of the fancier pieces of equipment is a laser for measuring the faintly scattered light from a layer of sulphate particles 70,000 feet up, made from volcanic gases, which is one of the permanent features of our atmosphere.

The most important single instrument at Mauna Loa has been recording carbon dioxide since 1958. The trend has been inexorably upwards, perhaps accelerating since the late 1960s. But of Reid Bryson's human volcano there is no sign at Mauna Loa; the windows of the greenhouse are not all dirty. Sensors that track the Sun and measure its brightness, at different slanting angles through the atmosphere, recorded a sudden increase in the interfering dust in 1963 – but that was from a real volcano, Mt Agung. When the dust of Agung settled, the dust reverted to its previous lesser abundance.

Yet palls of dust do hang over particular regions of the Earth, where soil erosion is bad, where cement is manufactured, or where cities pour out their smoky refuse into the air. While places downwind from large oceans, like Mauna Loa and the west coast of Ireland, show no long-term change in dustiness in the air, many other observatories in the northern hemisphere report increases, which reduce the sunshine reaching the ground. For example, combined measurements from eight stations in the USSR show that dust reduced the sunshine by ten per cent between 1940 and 1967. Japan's infamous air pollution has also had a marked effect, and near cities such as Mexico City and Jerusalem the man-made veil across the Sun is thickening rapidly. But the pattern is still patchy and the overall climatic effect is uncertain.

Sampling the air around the world shows nature's contribution. Clean air, in mid-ocean, has fewer than sixteen thousand particles per cubic inch although the bursting bubbles of the ocean surface put 10,000 million tons of salt into the air, as microscopic crystals. Over sparsely populated continents there are five times as many particles – either soil or organic materials released from plants as a natural smog. Where man makes his influence felt, in farmlands and suburbs, the particles multiply again by five. Over cities and centres of industry the particle count shoots up to more than $1\frac{1}{2}$ million per cubic inch.

The detector that counts the particles works by mak-

ing water vapour condense on them, to form a fog. And that is one way that dust particles affect the weather machine. Without a supply of particles on which water droplets and ice crystals can form, there would be no clouds. If the extra particles added by human beings help to make more clouds, or icier clouds, that could cool the Earth by reflecting sunshine off the cloud tops, depending, though, on where and when the clouds appear and how high they are. The dust itself can block the sunlight, yet particles that absorb the rays actually warm the air around them. As with carbon dioxide, assessing the real effects of man-made dust is not straightforward. Except in the stratosphere, rain continually washes away dust, cleansing the air thoroughly. But when the air is sinking, horrible smogs can develop from man's aerial refuse, in Los Angeles and Tokyo notoriously. Control of smoke in London reduced the fogs of winter by 85 per cent.

Like many other cities, St Louis, Missouri, alters the climate downwind from it. There, on a summer's day, you may see an instrumented aircraft pursuing clouds that form above an oil refinery, when there are no other clouds for miles around. Or it is taking samples of polluted air from above a chemical factory. Another aircraft flies into clouds far from the city, detecting the particles they contain. Sensitive radars monitor the clouds over a wide area, while meteorological balloons probe the conditions in the upper air. Research groups from six American laboratories are cooperating for five years in a Metropolitan Meteorological Experiment, METROMEX. They want to pin down exactly how St Louis affects the weather, as a guide to what may happen elsewhere.

Stanley Changnon of the Illinois Water Survey inspired this intensive study of a city and its weather. He selected St Louis, because there seemed to be, in summer, a large area of unusually heavy rainfall downwind

of the city. Now the scientists besiege the area with instruments each year. Already they have shown that half a million hectares downwind of the city have 30 per cent more rainfall than surrounding areas, and twice as many hailstorms.

The reasons are still ambiguous. The city apparently increases the chances that an existing cloud will shed its rain. The balloons and aircraft have mapped out a dome of warm, dry air over the city. It could very well cause passing clouds to weaken and shed their loads of ice and water. But the effect of the city is much less at the weekends, so industrial smoke and dust must be playing a part too.

Taking the offensive

The possibilities for human influences on weather and climate are legion: I could fill this book with them. One reason why I don't do so is the deep-rooted uncertainty among the experts. Here very briefly are some other examples, besides man-made dust and carbon dioxide.

Man-made heat. New York City generates seven times more heat than it receives from the Sun and the fact that cities are warmer than the surrounding countryside is well known to weathermen. Globally, at present, the human contribution of heat is very small. The energy supplied by the Sun is six thousand times greater than all our fires and engines and nuclear furnaces, and one thunderstorm can release more energy than New York does in a year. But human heat sources certainly affect local climates and probably regional climates as well. For example, urban concentrations in the eastern United States might tend to attract Atlantic storms to the coast. Sooner or later mankind's ever-increasing demands for energy may have to be checked for climatic reasons – unless, that is, we need all the energy we can make, to ward off an ice age.

On the lonely mountain an observer tends instruments that use the apparent brightness of the Sun to record the dustiness of the air. The Mauna Loa Observatory is the first of several monitoring stations for global as opposed to local pollution.

Attacking the clouds. The aircraft's burner releases fine 'seeds' of silver iodide, which encourage cold water droplets to turn to ice.

High-flying aircraft. Artificial clouds, in the form of condensation trails, are familiar sights. Higher up, jet engines release into the dry, thin air of the stratosphere water vapour, carbon dioxide and reactive compounds of nitrogen. One alarm has been that the reactive gases, especially from supersonic aircraft like Concorde and the Tupolev-144, could partially destroy the layer of ozone high in the stratosphere. The ozone keeps the stratosphere warm and also shields the ground from deadly ultraviolet rays of the Sun. Intensive research on stratospheric chemistry is in progress in several countries, especially the United States. But the results may be rendered inconclusive by a discovery in Canada that the natural reactive nitrogen gases in the stratosphere vary enormously from hour to hour.

Man-made deserts. For ten thousand years farmers have been destroying forests and other natural vegetation and sometimes dust bowls have resulted. By this test, though, locusts, plant diseases and forest fires should also be counted as agents of climatic change. Jule Charney's new theory of deserts (page 71) implies that destruction of vegetation encourages the air to sink, although he himself suspends judgment about whether man is to blame for enlarging deserts. More positively he suggests that efforts to revegetate the desert margins could well be rewarded with an increase in the local rainfall.

Dams. These create artificial lakes, often in desert areas. Dams also mitigate the flooding due to heavy rainfall. Downstream they may reduce the natural vegetation by starving it of water, or increase the planted vegetation in irrigation schemes. In theory the man-made lakes should tend to energise the air over which they pass and increase rainfall in the region. As far as I know there is no clear evidence of any such effect. Special anxiety among climate experts surrounds the Soviet scheme for diverting the great Siberian rivers

that flow into the Arctic Ocean in order to irrigate the increasingly arid lands of central Asia. The fear is that the loss of fresh water, which freezes readily, will reduce the sea ice in the Arctic, with unknown effects on the world's climate, but they might include worse droughts in central Asia and a profound cooling of the stormy zone.

These examples lead us to the possibility of deliberate human action to alter the climate – of going over to the offensive in the war against bad weather. Melting the Arctic sea ice is, in fact, one of the possible objectives that scientists speculate about. You might do it by spreading soot on the ice; or by exploding H-bombs to make clouds at the right height to warm the Arctic; or by damming the Bering Strait between Siberia and Alaska and pumping water into the Pacific to draw the Gulf Stream further north on the far side of the pole. The discoveries about 'thermal forcing' by the sea temperatures suggest that you might divert the jet stream by pumping up cold water from deep in the ocean, at selected places. Another proposal is to water the desert by drawing the moist air upwards in huge plastic chimneys, two miles high and 300 feet wide, supported by a balloon. You could put huge mirrors or metal vapours in orbit around the Earth to capture some of the Sun's rays that bypass the Earth. But all these ideas are fanciful at present – not just because of the technical and economic difficulties but because no one can say what the consequences would really be.

Apart from blasting away fog at airports by brute force (at Paris Orly they use 15 jet engines), the only substantial attempts to modify the weather at present aim at promoting rain, reducing hail or moderating storms by small-scale 'seeding' of clouds. Clouds appear when moist air rises and reaches the level where the air becomes too cold to keep its water vapour. But the water droplets and ice crystals of which clouds are

Lines of cloud formed in the wake of high-flying aircraft symbolise man's unintended effects on the atmosphere.

composed need those small particles – dust of different sorts – as nuclei on which to form. The sea-salt particles are good for making water droplets in comparatively warm air. Other particles promote ice formation in colder air higher in the cloud, and most of the world's rain is melted snow. In the 1930s, in the Netherlands and Germany, the idea of helping nature by supplying nuclei artificially was pursued but without success.

In 1946 two scientists working for General Electric in Schenectady pioneered cloud seeding. Vincent Schaefer promoted ice formation in clouds using dry ice (solid carbon dioxide) and then Bernard Vonnegut came up with silver iodide. This material, dispensed as a smoke of very small crystals from burners or from fireworks, is still the prime agent of the would-be weather modifiers. Silver iodide crystals closely resemble ice crystals and so act as excellent nuclei for converting ready-cooled water droplets into ice. Certainly silver iodide can alter clouds at temperatures below −5 degrees in a spectacular fashion, and onlookers are easily misled about its efficacy.

Among the weather-modification efforts around the world that are deemed to be operational rather than experimental, a notable example is South Dakota's. There control of the weather is enshrined in state law and, over the wheatfields of the High Plains, seeding aircraft feed smoke trails of silver iodide into the updraughts of clouds. They are guided to their targets by ground radar. Almost a million dollars is spent each year seeding clouds over half the state with the aim of altering the climate to the extent of giving the farmers an extra few inches of rain during the growing season. In addition the South Dakota Weather Control Commission offers to suppress hailstorms with fireworks shot from its aircraft. Most farmers believe it all works and happily pay three cents per acre for the service.

The farmers of Yugoslavia are happy too – in fact

Seeding affects clouds – to what avail is still controversial. 'Before and after' photographs, left, record a pair of puffy clouds transformed into an icier mass 16 minutes after Joanne Simpson and her colleagues seeded them with crystals of silver iodide.

Hailstorms can ruin crops (right). Soviet anti-aircraft shells (below) carry seeding agents into suspect clouds, to make more, but smaller, ice particles.

they join in the work of shooting rockets at hailstorms. Teams of professionals, equipped with radars, are charged with defending large areas of the farms and vineyards against hail, which can destroy a year's labour in a few minutes. But they command several dozen hail-control stations scattered through the countryside, each equipped with rockets and manned by local farm workers. When a suspect cloud seems, by radar, to be capable of dropping quarter-inch hailstones, orders go out to aim some rockets at the level in the cloud where water droplets, overdue for freezing, are accumulating. The technique follows faithfully the methods of hail-control devised by Soviet meteorologists. Each rocket strews the cloud with silver iodide smoke. The idea is to encourage the cloud to make many small particles of ice instead of the fewer but heavier hailstones that cause the damage.

The Soviet meteorologists have claimed astounding successes – with four-fifths of damage due to hail eliminated by this method. Most foreign experts are sceptical about that claim, although several countries are now testing the Soviet methods. The rocket-rattling farmers of Yugoslavia are convinced of their own successes. But that is the trouble with trying to control the weather: how do you know that the cloud you shot at would really have dropped hail if you had not fired?

The most outspoken critic of the weather modifiers is John Mason, the director-general of Britain's Meteorological Office, and himself a distinguished cloud physicist. He argues, in essence, that until you can predict the natural behaviour of clouds you are not entitled to claim that you have changed it. Meanwhile, to evaluate claims of rainmaking you have to compare statistics, over long periods, showing what happened when you seeded and when you did not; but the more carefully that is done, the less impressive the results become. Mason therefore regards rainmaking as scien-

tifically unproven. He is also worried about the likely disputes between nations about their weather-control policies, if modification became a reality.

The basic issue, then, remains controversial, whether human beings can, by seeding, alter the behaviour of clouds to a worthwhile extent. In a prolonged experiment over Florida for the US government's Environmental Research Laboratories, Joanne Simpson and William Woodley are trying to settle the issue. They predict how clouds will grow. They predict whether seeding will affect the clouds' growth. On days judged suitable for seeding they fly into selected clouds, releasing silver iodide flares. But when they give the order, only the man responsible for the mechanism that releases the flare knows whether it really goes or not. He follows the instruction of a randomly drawn card that says 'seed' or 'no-seed'. Such are the lengths that Simpson and Woodley go to to avoid unconscious bias in their choice of clouds, their decisions to seed or their judgments of how the clouds changed. And although they have been running this kind of experiment, off and on, since 1967, they don't expect to have enough statistics for a final verdict on the effects of seeding until 1976.

Whatever the verdict may be – and even this careful experiment cannot satisfy everyone – cloud seeding by silver iodide is undoubtedly the right *kind* of idea, if man is ever to control the climate. It is a trigger, exploit-

ing the tendency of the atmosphere to produce effects that far outstrip the cause, in this case by releasing the energy and moisture already latent in the droplets that are overdue for freezing. Once the energy has been released it can, in principle, create an updraught that makes the cloud go on growing in a more natural fashion.

Are typhoons necessary?

On a wet afternoon at the Philippines Air Force base outside Manila, the colonels spoke of their plans to fight typhoons – the greatest storms on Earth. They showed me first the four-seater Cessna 210 aircraft, which are kept busy seeding clouds to 'supplement the nation's water resources' by rainmaking. They could hardly take on a typhoon. Then the colonels pointed to a big C-130 (Hercules) aircraft standing on the runway. From the Americans, the Marcos government is buying two such aircraft to be in action by 1975, thoroughly fitted out with meteorological instruments. Even more to the point, the Filipinos have 'bought' a belief, of which the C-130s will be the embodiment: the belief that typhoons can be tamed. In 1973 a Typhoon Moderation Council was set up by presidential decree.

Of all the world's major countries, the Philippines Republic is the most battered by tropical storms because it sprawls across the main highway of the Pacific

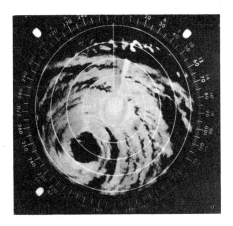

typhoons. In an average year nineteen of them hit one or other of the major islands comprising the nation. In December 1944, just off the Philippines, a typhoon intervened in the Pacific war, catching a huge American fleet in the midst of refuelling at sea. The raging water swept the flight decks of the carriers, made battleships heave to, and sank three destroyers with all hands. In September, October and November 1970, a succession of four typhoons struck the islands unusually hard and killed 1500 people. Patsy, the last of the four, hit Manila. Since then, the civilian weather service of the Philippines has accelerated its research into typhoons, and the Typhoon Moderation Council is now looking to the Americans for assistance – after all, the US Project Stormfury was what raised the Filipinos' hopes. If man is to put the world's climate to rights, is this not a good place to begin?

About once a week, somewhere in the tropics, a straggling cloud cluster begins to arrange itself in a sinister fashion. Like a tadpole changing into a frog, the cloud cluster takes on a new appearance: a tightening ring of clouds from which curved cloudy arms begin to extend. The weathermen checking the satellite pictures hurry to issue warnings. And they look through the coy lists of girls' names to discover that this new monster is to be designated Hurricane Dolly, perhaps, or Typhoon Dolores.

The great storms are called typhoons in the North Pacific, hurricanes in the Atlantic and cyclones in the Indian Ocean and Australia. They are all one species, the tropical cyclones, but the most savage are to be found among the Pacific typhoons and the deadliest among the Indian Ocean cyclones. The naming is further confused because 'tropical cyclone' often refers to storms much weaker than typhoons or hurricanes. Here I am dealing with the big ones.

They are creatures of the tropical oceans. Conditions have to be just right, before the scattered thunderstorms of a tropical cloud cluster can turn themselves into a cyclone. The surface water of the sea has to be very warm, more than 26 degrees; a tropical cyclone cannot form very far from the equator. Neither can it form too close. Only where the globe begins to slope inwards towards the poles does the spinning of the Earth begin to grip the air and organise it into a vigorous eddy. That is why tropical cyclones are commonest in summer and autumn, when the Sun has had a chance to heat the sea well to the north or south of the equator. Even so, only about one in ten of 'promising' cloud clusters becomes a tropical cyclone: the engine needs a starter motor in the form of suction supplied by the winds of the upper air. A better grasp of this process is expected from the 1974 GATE experiment, which was staged in the hurricane spawning-ground of the tropical Atlantic.

The storms normally head westwards; sometimes they 'recurve' sharply away from the equator and plunge across cooler waters. Others follow crazily looped tracks. Countries on the west of the oceans are most at risk. Australia (where Brisbane was assailed in 1974), the Philippines, China and Japan are open to the Pacific storms; the Caribbean islands, Mexico and the south-eastern United States to the Atlantic hurricanes. Sometimes a hurricane will completely recurve and head towards Europe, but by the time it arrives it will have degenerated into an ordinary depression. Although 'hurricane-force winds' of more than 64 knots may be registered at times almost anywhere in the world, to qualify as a true hurricane the storm must possess that compact whirliness which makes 'cyclone' the most descriptive of the various names. A mature tropical cyclone leaves an unmistakable signature on a satellite picture, strangely beautiful, like a spiral galaxy seen through an astronomer's telescope.

Typhoon Ida, 1958, brought far too much rain to Tokyo – yet Japanese experts are mistrustful of proposals to moderate typhoons.

The storms often massacre ships and boats. One hurricane off Hispaniola in early July 1502 fell upon a fleet of thirty Spanish ships homeward bound with ill-gotten gold; twenty of them sank outright with their crews and treasure and only one reached Spain. On land, the wind can flatten crops, uproot trees and blow flimsy houses away, but floods caused by the heavy rain are more destructive. Worse still, the combined effects of atmospheric suction and geography can heap up salt water in a storm surge and pour it over the land. Bays with shallow seabeds are the most vulnerable places. For instance, more than 5000 Japanese lost their lives in a typhoon that lifted the water of Ise Bay ashore in 1959; in 1900 6000 Texans were drowned by the sea during a hurricane at Galveston, but that may not have been strictly a storm surge.

Storm surges are the great killers and nowhere more than in the Ganges-Brahmaputra delta at the head of the Bay of Bengal. In October 1737 a cyclone lifted the sea by more than 20 feet and drowned 300,000 people. Exactly the same thing happened in November 1970, only this time the whole world knew it was coming except for the people on the spot. The storm showed clearly on the satellite pictures and on the weather radar at Cox's Bazar. After a succession of lesser disasters in the 1960s (about 50,000 dead, in aggregate) and a stern warning from the former head of the US National Hurricane Center that many more might perish, an evacuation had been rehearsed. But the real trouble was a lack of local embankments where families in isolated communities might have scurried to safety until the sea retreated. In any case, no evacuation order was given and once again about 300,000 people died. So it goes.

At the heart of a tropical cyclone is an 'eye' about 15 miles wide and almost perfectly circular; the Japanese found one typhoon with an oval eye in 1966, but that was a rarity. Inside the eye there are light winds and

The air-and-water engine of a typhoon or hurricane. A cross-section of the storm (below), matched to the eye of a typhoon as seen from a satellite, shows moist air spiralling in towards the centre. It rises in the eye wall and spills out from the top. The greatest danger to life is the storm surge, as produced (right) by Typhoon Vera at Nagoya, Japan, 1959.

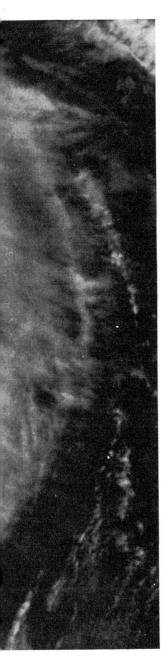

some low clouds. Often clouds high above the eye veil it from the watching satellites. But from an aircraft flying into the eye, as American hurricane-hunters do routinely, one sees a coliseum of mighty clouds surrounding the eye, and towering 35,000 to 40,000 feet high. This 'eyewall' is the chief engine of the storm, the vortex where the wind screams loudest as it whirls around the eye and whips the ocean water into a blinding cream of spray. Wind speeds of over 100 knots are quite usual.

An updraught of air in the wall helps to drive the whole storm. It creates a partial vacuum; the bottom seems to fall out of the barometer and the storm gulps in air from hundreds of miles around. The air blows in hot and damp over the ocean along curling paths that gradually bring it towards the eye. Charged with energy stolen from the ocean, the air makes many small clouds, exchanging some of its water vapour for heat. Then, in the wall, it makes the great leap skywards, shedding the remainder of its water. Over the top of the storm, the cooled and dried air spills outwards; only a little of it subsides gently into the eye itself.

Enquiries into how tropical cyclones work now depend upon numerical models that vent their fury and pour out their imaginary rain in the circuits of a computer. So far the model cyclones approximate only roughly to the real things, but they are improving and the value of putting numbers to all the processes is that the researchers can begin to see which features of the storm are cardinal and which are by-products. For example, older theories paid too much attention to the release of heat in the tall clouds of the eyewall and not enough to the contributions from all the lesser clouds formed in the spiral bands of the inrushing air. The characteristic pattern of spiral rainbands seems to be formed by a wave in the air which alternately encourages the formation of clouds.

A cyclone lives much longer than does an ordinary cloud cluster. Typically it has about 10 days in which to run berserk. Once formed it is free to travel outside the zone of hot, tropical sea water – but then it is bound eventually to weaken, as it must also do if it invades a continent. It can be stifled, too, by strong or confluent winds in the upper air, which are commonest outside the tropics. But often a cyclone passes over land or a patch of cool sea water, and then finds warmer water where it can feed again and recover its full vigour. The chief problem for the forecaster presented with a satellite picture of a new hurricane or typhoon is to predict its track, and especially any landfall it may make, with the storm surge as the greatest worry. Most storms most of the time proceed steadily (at perhaps 30 knots) along a slightly curved track; the challenge is to predict abrupt changes of course or speed which could bring disaster to an unprepared district. The forecaster cannot cry wolf very often before people will disregard a warning that really matters.

At the National Hurricane Center in Miami, American meteorologists use four different techniques for predicting the track, and they try to pick the most appropriate technique for the storm in question. The simplest one assumes that the hurricane will persist in its present motion unless general knowledge of the behaviour of hurricanes at this time of year suggests otherwise. The favourite method uses all available records of hurricanes closely resembling the present one, and combines them to show the most likely positions over the next three days. Anticipating a northward movement by the hurricane is often a matter of the utmost importance to the USA. A third approach is best for this purpose: the forecasters consider the weather over a large area and see how it may affect the track. Finally the winds around the storm are modelled in a computer in a rather abstract fashion to find out where the vortex will shift; this technique is worse than the others for the first 24 hours but better for two or three days ahead.

A typhoon hitting Hong Kong in 1937 killed 11,000 people. A more severe one in 1962 made 72,000 people homeless and destroyed many crops and 1300 boats; the loss of life was 130, scarcely more than a hundredth of the 1937 disaster's. Except in underprivileged areas like the Ganges delta, satellites and radars, predictions of storm surges and public warnings by radio and television have (touch wood) enormously reduced the death-toll of tropical cyclones. But the outcome is largely decided long before the storm brews – by building windproof houses or storm shelters, by flood gauges and dams on rivers, by special sea walls like the one I have seen in Osaka, Japan, which can hold back a storm surge ten feet high. Sometimes the needs of safety conflict with good taste. The Florida coast will be safer for its inhabitants and visitors when robust, high-rise hotels and apartment blocks replace the little houses and motels; then, when the hurricane brings its storm surge you carry your martini upstairs. (Tell that to the Bengalis!) But even as the loss of life is curbed, the material damage done by a tropical cyclone increases, as it finds expensive trappings of modern civilisation to toy with: power lines, telephone cables, cranes, airfields, advertising signs, scaffolding, radio antennas, cars and trailers, and so on.

So the argument for taming tropical cyclones is economic rather than humanitarian. When Project Stormfury started in the United States in 1962, as a joint effort of the US Navy and Department of Commerce, the mood of the times favoured the 'technological fix' for the world's ills. It seemed entirely reasonable to confront the impertinent and exorbitant hurricanes with human ingenuity. Would anyone who succeeded in defeating the storms not deserve well of mankind?

Two main notions about how to do it have persisted over the years. One is to cut off the storm's fuel supply – the water vapour from the ocean – by painting the sea with a chemical film. The drawback is likely to be the storm's manifest capacity for stirring the ocean surface. The chief idea in Project Stormfury is to seed the cyclone with silver iodide smoke on the outside edge of the eyewall. The object is to release heat and build clouds that will spread out the zone of high winds, so weakening them and the whole storm. Rough tests were carried out in 1961 (Hurricane Esther) and 1963 (Beulah) but the principal experiment was in 1969. Aircraft seeded Hurricane Debbie on 18 August and the maximum wind (at 12,000 feet) dropped from 98 to 68 knots; during the following day the storm recovered and on the third day, 20 August, a repetition of the seeding reduced the wind from 99 to 84 knots. It was encouraging, to say the least. Cecil Gentry, director of the National Hurricane Research Laboratory in Miami, told me that when he took the job in 1966 he thought the odds against modifying hurricanes were long – now he thinks there's an even chance of success. But the credibility of the technique depends in part on the supply of water droplets overdue for freezing, ready to turn into ice when the seeding is done.

In September 1973, Hurricane Ellen steamed northwards across the Atlantic east of Bermuda. Two American research aircraft flew into it, on behalf of the National Hurricane Research Laboratory. One of them, a C-130, was the most thoroughly equipped aircraft ever to penetrate a hurricane; there were six different instruments for measuring water droplets and ice particles. Sinkable thermometers dropped from the aircraft gave evidence of deep-lying waves in the sea in the wake of the hurricane. But most important for the Stormfury theory was the discovery of even greater quantities of liquid water in the clouds, at temperatures below the freezing point, than the experimenters had hoped to find.

But by then Stormfury was in temporary decline. The US Navy pulled out of the project in 1972 and the civilian funds were going largely into preparing new aircraft and instruments. And because Atlantic hurricanes are comparatively rare, the project was adjourning to Guam in the western Pacific, where typhoons abound and seeding would resume in 1976. Meanwhile two kinds of meteorological critics confront the would-be stormbusters.

Some raise technical objections that the methods proposed either could not affect the storm or, if anything, would make the winds stronger – as you might expect from a process that adds to the storm's overall release of energy. These critics explain the apparent success with the seeding of Hurricane Debbie in 1969 as due to natural diminutions of the storm, perhaps because it ran over cool patches of sea water. Others point to the enormous natural release of energy in a fully fledged hurricane, equivalent to an H-bomb going off once a minute. Is it really going to notice a few pounds' weight of silver iodide? While further experiments are delayed, the controversy is waged with rival computer-calculated hurricanes. The trouble is that, so far, none of the numerical models represents the physical processes in the clouds in a very realistic fashion.

The other kind of objection is more fundamental and comes most forcefully from the Japanese. They suffer four typhoons a year, on the average, and the damage done by the average typhoon costs them $100 million. The early-warning weather radar erected on the summit of Mount Fuji by a prodigious feat of engineering is a monument to Japanese concern about typhoons. From its vantage point 12,000 feet high, it can spot the tell-tale spiral arms of a typhoon 50 miles offshore. A chain of sixteen lesser radars tracks storms approaching particular segments of the coast and a lot of effort goes into preparedness, warning and rescue. But the ill wind blows a lot of good to the farmers. A quarter of Japan's rainfall comes with the typhoons. The Japanese are quite prepared to put up with their typhoons and are very suspicious of anyone who wants to play games with them. They are insisting that the forthcoming Stormfury operations in the Pacific be restricted to typhoons that could not hit land. The Filipinos, in the face of the Japanese misgivings, are beginning to talk of their typhoon moderation programme as a research effort rather than an operational one; or perhaps they too suspect that seeding might make a typhoon worse.

The tropical cyclones may be an essential component in the weather machine of the whole world as it functions at present. If you could really smash the Atlantic hurricanes, for example, you might reduce the rainfall in Europe. Individual hurricanes often turn into, or link up with, rain-bearing depressions travelling in that direction. More basically, tropical cyclones are a maverick part of the transport of warmth from the tropics towards the poles, which drives the winds of the world and helps to hold back the Arctic ice. If typhoons and hurricanes are to be controlled at all, it may be wiser to try to steer them away from densely populated areas rather than to moderate their force.

If the Earth's climate goes on growing cooler, typhoons and hurricanes may become rarer in any case. Japan and the Caribbean nations may then be complaining of drought, and looking for ways of promoting the storms. The moral of all this is that weather modification could easily become a contentious issue in international politics, because one man's shipwreck is another man's harvest. It does not augur well for concerted human action to try to control the climate and avert disastrous changes, even assuming that we knew how to do it.

A computer's view of the Earth and its climate. The lower map shows average surface temperatures calculated by the joint ocean-atmosphere model of Syukuro Manabe and Kirk Bryan at the Geophysical Fluid Dynamics Laboratory, Princeton. The upper map of actual surface temperatures shows a reasonably good match. But the performance of computer models will have to improve markedly before they can predict subtle changes of climate. (Absolute temperature: 273° = 0° centigrade.)

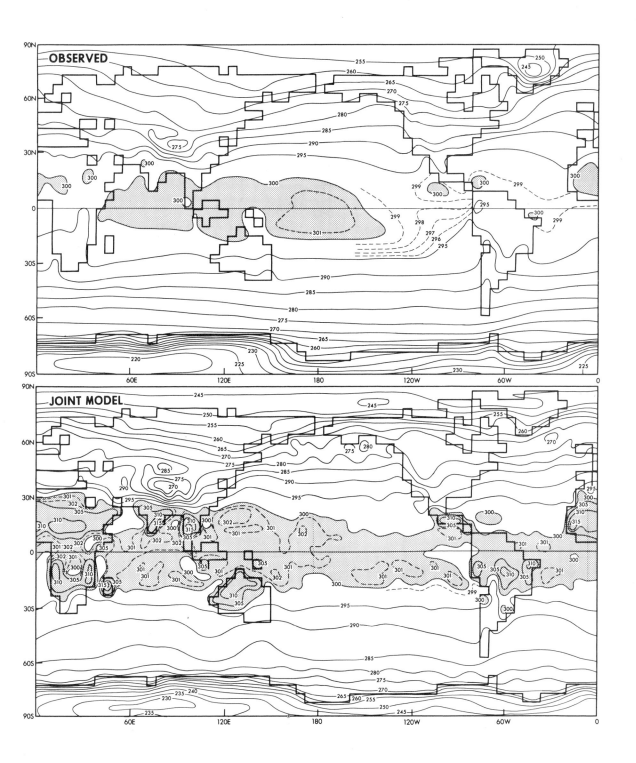

Climate by numbers

At the Geophysical Fluid Dynamics Laboratory in Princeton, clustered around a big new computer like bears around honey, are some of the outstanding meteorological theorists of our time. One uses the machine to consider how cities may affect the weather or to see how pollution disperses in the Earth's atmosphere. Others compare the climates of other planets with the Earth's, or sum up the awful violence of a hurricane. Yet another is engaged in the boldest attempt yet in extending the range of numerical weather forecasts, aiming at 30 days. These researchers include a significant fraction of the Japanese population of Princeton, because Japan has produced an exceptional crop of meteorological theorists, but the Americans have the computers they need.

The machine at Princeton can carry out a hundred operations in a millionth of a second. A well-tried numerical model computes the atmosphere at 18 different levels. Given these tools, Syukuro Manabe wants to describe the Earth's climate in the computer well enough to explain past climatic changes – and then to predict future changes. Earlier, I referred to his work on the monsoons.

The weather machine is a system of three fluids, air, water and ice, kept in continual motion by the Sun's heat and always interacting with one another. With an oceanographer, Kirk Bryan, Manabe has made a model atmosphere interact with a model ocean. The combined model includes provision for snow and sea ice to form and melt; evaporation and rainfall, and the fate of rainwater on the continents, are also taken into account. But they have had to make some unreal simplifications; abolishing the difference between winter and summer, for instance, and telling the computer in advance what was to be the distribution of clouds and gases affecting radiation to and from the model Earth.

Because the atmosphere changes so much faster than the ocean, they let the model atmosphere run for a model 'year' to show its general behaviour; then the model ocean runs for 300 model 'years' in response to those atmospheric conditions; then the atmosphere runs again in accordance with the changed oceanic conditions – and so on. Considering all the simplifications, the patterns of temperature, rain and ice are quite realistic. For example, the combined model has captured well the role of the ocean currents in shifting the rainfall belts of the stormy zone further north or south from the equator than they otherwise would be; also the reduction of rainfall at the equator where cold water wells up from deep in the ocean.

Perhaps the biggest single shortcoming is the failure to model changing patterns of cloudiness. The clouds threaten to be the Achilles heel of numerical models of climate. A surprising discovery by satellite illustrates the importance of clouds. The southern hemisphere absorbs no more and no less heat from the Sun than does the northern hemisphere, despite the immense preponderance of ocean in the south. Only a massive correction to meteorologists' expectations about the cloudiness of the southern hemisphere can account for the balance. One of the difficulties in dealing with cloudiness in numerical models is technical, having to do with the mass of the clouds being minute while their effects are great. But there is also too little knowledge of the typical behaviour of the different kinds of clouds in different settings. For instance, high clouds in the tropics have a warming effect below, while in other places they have a cooling effect.

There may also be uncertainties about the realism of the ocean model, especially in view of the recent discoveries of the eddies, which are too small to be represented in the model. In fact this pioneering effort by Manabe and Bryan, impressive though it is, shows how

Vegetation as an indicator of climate – here, the open 'Páramo' on the mountains of Colombia in South America. During the cooling and droughts coinciding with ice ages, vegetation of this kind displaced large areas of tropical forests.

The icy part of the weather machine. Ice like this, at the Arctic island of Spitzbergen, greatly influences even the present-day global weather. Intermittently the ice seizes huge areas of the northern continents.

difficult it may be to achieve the realism and precision needed to explore cause and effect in small changes of climate, amid the great variability of the weather that occurs even without climatic change. In any case there is always a limit to the realism attainable in a reasonable period of computing time.

The other chief numerical models of climate have originated at the National Center for Atmospheric Research at Boulder, the University of California at Los Angeles and the Meteorological Office at Bracknell. All the theorists working with them face the same kinds of problems and, although Princeton has the most computing power, even that seems inadequate for the progress that will be needed. The Princeton machine is 10,000 times faster than the computers of two decades ago, but the end may be in sight: at 100 times faster still, the speed of light may limit computer speeds.

There will be little hope of simulating, by computer, the interactions of air, water and ice over the thousands of years involved in the onset and relief of an ice age, without some far-reaching progress in the methods. One idea is to treat the great disturbances of the stormy belt, which are very time-consuming for computers, by statistical rules for their behaviour. That could be self-defeating, though, if the key to climatic change were a shift in the behaviour of the storms. Another approach may be more general arguments about key factors to deduce possible conditions of the weather machine and to use the numerical models to see whether those conditions could really arise and persist.

Joseph Smagorinsky, a pioneer of numerical models of the atmosphere, who now heads the Princeton laboratory, is convinced that only by numbers can one resolve the confusing interactions of cause and effect among the many different pieces of the weather machine. He thinks that crude or premature estimates of climatic effects of human activity, for example, can be

worse than no estimate at all, and that we should be wary of basing national or international decisions on 'handwaving arguments or back-of-the-envelope calculations'. As he remarks:

These new anxieties regarding the possible impact of changes in climate, either natural or due to man-made influences, are coming at a rather fortunate time in the development of meteorological science. Laboratories throughout the world are busily engaged now in constructing models – that is, developing rules by which very large computers can make predictions about the future evolution of the atmosphere, the oceans, and the ice interface. Although these models are rather simple now and their predictive capability can only go over very short periods of time, we hope that eventually they will give us a capability of saying what will happen over longer climatic eras.

But some investigators of climatic change are sceptical about the computers; Hubert Lamb, for instance:

The role of the numerical modellers for a long time to come is, I think, likely to be in helping towards broad physical understanding of the mechanisms, large and small, involved in changes of weather and climate. This physical understanding will greatly help in the finding and formulation of statistical forecasting rules. There are probably too many complexities and indeterminacies of detail for the numerical model ever to produce the actual forecast for the periods concerned in seasons, years and decades ahead.

Just as 'empirical' methods of forecasting, based on factors like sea temperatures and comparisons with previous years, at present outstrip the numerical models for 30-day forecasts, so Lamb thinks that comparable methods will be needed for climate forecasts.

The numerical modellers, with their computers, confront the intricacies of the weather machine. They doubt the rather vague meteorological arguments of the investigators of past climates. Those investigators in turn doubt the ability of computers to capture the slow and subtle changes in the weather machine, for which they have concrete evidence from the past. Almost certainly these points of view will have to be reconciled. Meanwhile the divisions are debilitating. They have become

scandalous in Britain, where Lamb's own climatic research unit, esteemed throughout the world, has been
denied government funding owing to opposition from
the meteorological establishment.

As we shall see in the next chapter, a numerical
model is reconstructing the weather of the most recent
ice age, on the basis of information about sea temperatures and ice sheets deduced by 'empirical' climatologists. That collaborative experiment may signpost the
way ahead, as human beings try, for the very first time,
to guess what the weather will be like ten years or a
hundred years from now. And, because they illuminate
the link between increasing cold in the north and
drought in the tropics, ice-age studies are not irrelevant
to the present, lesser climatic changes.

Pending the greater understanding that the modellers
and the empiricists both crave, a judicious verdict on
what is happening to the weather of our time must be
that there is cause for fear. In many parts of the world,
from the Arctic to the tropics, harvests are failing
where previously the yields were reliable. The simplest
and most likely reason is that the early part of this
century represented a short break in the Little Ice Age,
which is now resuming.

The chief contrary hope must be that the cooling in
the north that has proceeded since 1950 will reverse.
The northern hemisphere in 1973 seems to have been
a little warmer than in 1972. Only time will tell whether
this change, like the recent mild winters in northern
Europe and eastern North America, marks a reversal in
the trend or merely a temporary deception.

*Time-scales of climate: the darker the tint, the colder the condi-
tions. We live in one of the Earth's icy periods (first three
scales), but in one of the warm periods between ice ages
(130,000-year scale). Seen on the 10,000-year scale, though, we
are in a comparatively cool 'little ice age' of the kind that recurs
every 2500 years. The last scale shows us apparently entering
one of the cooler phases of our Little Ice Age, after a warm spell
earlier in this century.*

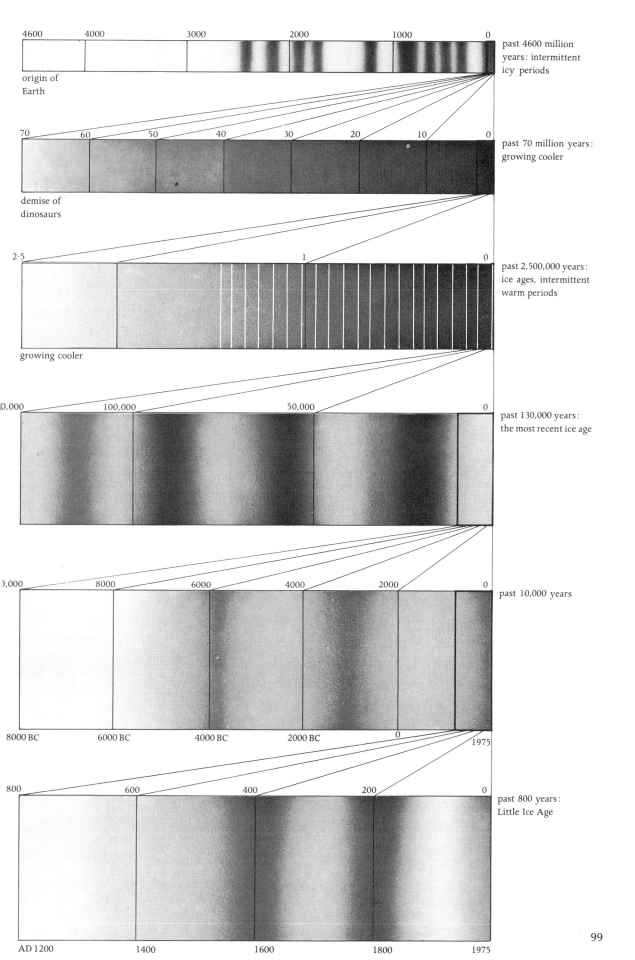

4600   4000   3000   2000   1000   0

past 4600 million years: intermittent icy periods

origin of Earth

70   60   50   40   30   20   10   0

past 70 million years: growing cooler

demise of dinosaurs

2·5   1   0

past 2,500,000 years: ice ages, intermittent warm periods

growing cooler

0,000   100,000   50,000   0

past 130,000 years: the most recent ice age

0,000   8000   6000   4000   2000   0

past 10,000 years

8000 BC   6000 BC   4000 BC   2000 BC   0   1975

800   600   400   200   0

past 800 years: Little Ice Age

AD 1200   1400   1600   1800   1975

99

A piece of deep-lying ice from Greenland tells Willi Dansgaard of a drastic event long ago.

3 The Threat of Ice

A curious feature gradually appeared in the Greenland ice as the oxygen analysts in Willi Dansgaard's Copenhagen laboratory worked their way through the very long core that had been drilled in 1966 at Camp Century in northern Greenland. It was in ice from very deep down, near the bottom of the ice sheet. This represented snow that fell during the warm period before the start of the most recent ice age. At first the feature showed in their diagrams as a little nick of cold, indicated by a decrease in heavy oxygen in the ice, at a date estimated at about 90,000 years ago. It was recorded in only a couple of feet of ice, out of the 4500 feet of the whole core from Camp Century. But more thorough testing of the ice that contained it revealed it to be more than a nick. The heavy oxygen content of the ice went very low indeed. By the early 1970s, Dansgaard and his colleagues realised the full import of the event.

The Earth of 90,000 years ago was jogging along in a comparatively warm period, not very different from our own. Quite dramatically, and seemingly in less than 100 years, the Earth plunged into extremely cold conditions equivalent to the full severity of an ice age. It was not, though, a typical ice age lasting for many thousands of years. The Earth climbed back to a warm climate in one thousand years. Dansgaard's sudden cooling might have been dismissed as spurious, or as wrongly dated or interpreted, or as a purely local event. Certainly the Danish group was stretching its techniques to the limit, or beyond, especially as far as dating was concerned. But the same event showed up in other parts of the world.

Compared with the Greenland ice, nothing else that comes readily to hand can be expected to show the full speed of so rapid a cooling. Earthworms scramble the layers of old soil on land. The record in the ocean bed is smudged because mud and corpses settle so slowly there; an inch in 2000 years is a fast rate of growth for the layers. Nevertheless, American fossil-hunters scrutinising samples from the ocean floor off Mexico have found an abrupt change in climate occurring in 'less than 350 years'. Species of small animals that lived near the ocean surface and liked warm water suddenly disappear from the 90,000-year level. Others that liked cooler water take their places. It was a revolutionary change in the life of the Gulf of Mexico, and the new mixture of species held the stage for 1000 to 1400 years.

And 90,000 years ago flourishing oak forests both in Macedonia (northern Greece) and in the Netherlands were wiped out within about a thousand years. That is known from Dutch research on fossil pollen, found in borings on land. French scientists reported oxygen measurements in samples from a stalagmite. It is a splendid column, eight feet high and 130,000 years old, growing steadily from dripping water two miles in from the entrance of a cave at Orgnac, near the mouth of the Rhône. These samples also showed a sharp drop in temperature in less than a thousand years, though starting 97,000 years ago according to the investigators. It may turn out to be the same sudden cooling, if the dates can be reconciled. Finding traces of this cruel event is a new game for fossil-hunters. And what happened to our kin, the primitive representatives of *Homo sapiens* but not yet of modern man, who were living in Europe at the time of the sudden cooling? Future archaeologists will no doubt be anxious to find out.

Hints of other sudden, brief periods of intense cold have still to be pursued in the Greenland ice record and the ocean-bed cores. But one case is enough to show how very abruptly the weather machine can change, all the way down through its range of gears and up again. Casting around for possible explanations for the sudden cooling of 90,000 years ago, the scientists see only two obvious mechanisms that might be capable of acting with sufficient speed. One would be an enormous volcanic eruption, or sequence of eruptions. But if it were thick enough to filter the sunlight and cool the Earth so precipitately, the volcanic dust should show up plainly in the Greenland ice or the ocean-bed deposits. In the Mexican ocean-bed samples that record the sudden cooling there is indeed a well-known layer of volcanic dust – but it comes a thousand years after the sudden cooling, not before it, so it cannot be counted as the cause.

The recently fashionable theory of the Antarctic ice surge offers another possible explanation for the sudden cooling. According to this theory the huge ice sheets that cover Antarctica even in our present fairly warm period can act like a climatic time-bomb. As they gradually build up, from the snow that falls on them, the pressure at the bottom of the ice reaches a point where the ice melts. Until that moment the ice has been able to travel over the ground, if at all, by deforming itself in the very slow process of 'plastic flow'. But once its base is lubricated by water, the ice can slide bodily, and much more rapidly. The friction of the ground now helps rather than hinders the progress of the ice, by causing further melting at the base.

Anyone who visits the ice sheets of Antarctica nowadays is likely to be haunted by the thought of what would happen if the ice were to slide into the ocean, taking perhaps a century or so. The notion has been much discussed. The ice would set up huge 'tidal waves' and the sea-level would rise 180 feet or more, drowning huge areas of the presently inhabited world. But the ice spreading out over the southern oceans could also cool the whole Earth drastically and trigger off an ice age in the north. Slowly, during the ice age and after, the ice sheet would form again, until the time came for the next surge. According to this ice-surge theory of ice ages, the Antarctic ice sheets have to accumulate for perhaps a thousand centuries before they are weighty

enough to melt the bottom layers.

Small-scale ice surges are a fact. The ice cap of Spitzbergen surged 13 miles in the late 1930s. Glaciers in Alaska have been known to stand almost still for half a century and then move downhill at three feet an hour. What is at issue is whether surges occur in Antarctica in a way that could make them the cause of the recurrent ice ages, or of a sudden cooling. At present the Antarctic ice creeps in a stately fashion into the sea, making icebergs at a rate roughly in keeping with the accumulation of new snow.

The Transantarctic Mountains effectively divide Antarctica: apart from some leakage through the valleys, the greater part of the Antarctic ice sheet is essentially cut off in East Antarctica. That ice sheet, by far the greatest in the world today, appears to be stable. Much of the ice it contains is far older than the most recent ice age. Any large surge of that East Antarctic ice sheet would have produced a sudden large rise in sea level all over the world; so far there is no clear evidence for such an event. A huge mass of spreading ice should also have left its mark in altered deposits on the ocean floor a thousand miles or more from the Antarctic coastline, because it would change the ocean currents and the distribution of plants and animals living at the sea surface. Closer examination of samples from the ocean floor may finally refute the idea of really large surges from East Antarctica. West Antarctica, on the other side of the Transantarctic Mountains and with a much smaller ice sheet, is a different case.

Due south of New Zealand is the Ross Sea, the deepest indent in the coastline of Antarctica and one of the chief escape routes for the ice of West Antarctica. Here the ocean reaches closest to the South Pole but, as James Clark Ross found when he first explored it in 1841, a huge barrier of ice stretches across the sea, flat-topped and reaching 100 feet above the surface. 'We might

The edge of the Ross Ice Shelf shows plainly in a satellite picture for New Year's Day, 1974. Some scientists fear that this Antarctic ice could surge into the ocean and so alter the sea level and climate throughout the world.

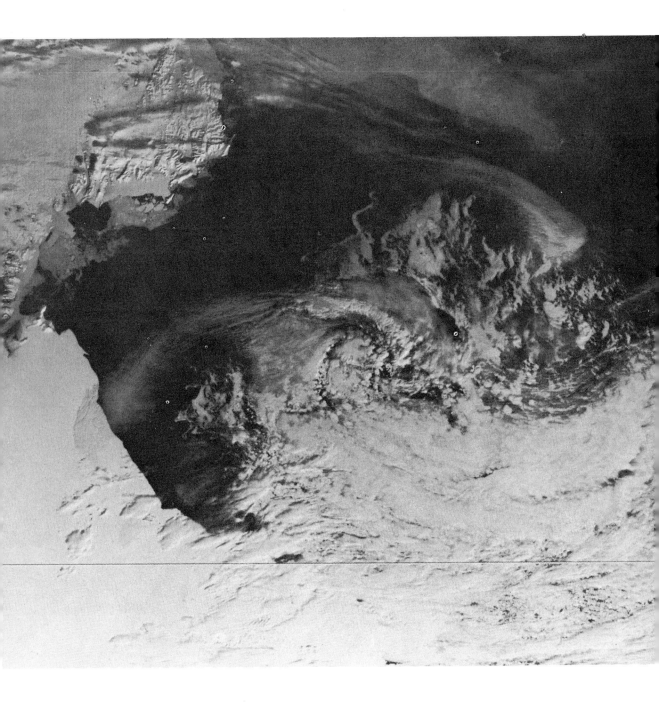

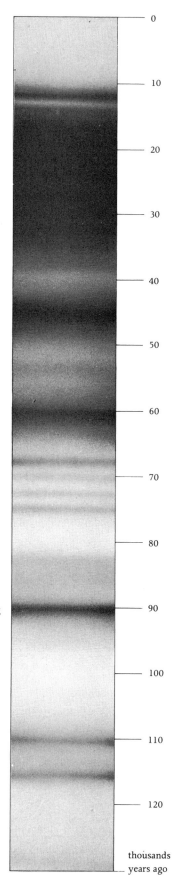

The cooling of 90,000 years ago, before the ice age proper, is pinpointed in the record from the ice of Camp Century. (Light parts of the column correspond to warm periods, dark to very cold periods.)

sudden cooling

0
10
20
30
40
50
60
70
80
90
100
110
120

thousands years ago

with equal chance of success try to sail through the cliffs of Dover,' Ross reported. The Ross Ice Shelf is now known to be a floating mass of ice the size of France. Fed by snow falling on it, and by ice creeping out of West Antarctica, the Shelf advances north across the Ross Sea at a speed of half a mile a year. The bottom of the ice gradually melts and the front of the ice breaks into huge icebergs which drift away, so the Shelf does not grow. But here is the most obvious place in the world for a major ice surge to occur. The picture would be of the ice of the Shelf marching seawards at a rate of ten miles a year, faster than it can melt.

Ice scientists have recently been busy on the West Antarctic ice sheet and the Ross Ice Shelf, trying to make sense of their behaviour. West Antarctica carries less than one-sixth of all the ice of Antarctica and its ice sheet is much more variable than East Antarctica's. Much of the ground underlying the West Antarctic ice sheet is below sea level, so that the ice sheet could vanish if the ocean became substantially warmer. Changes in the level of the sea, brought about by the freezing and melting of ice anywhere in the world, complicate the local behaviour of the ice. For example, if the sea level falls, the Ross Ice Shelf runs aground; if the sea level rises, more of the ice on the shore begins to float.

As George Denton of the University of Maine has put it: 'The glacier ice contained in the Antarctic ice sheet poses a potential threat to our environment because of its possible influence on sea level, oceanic circulation and climate.' Denton has revised much of the history of glaciers in Alaska and Sweden as well as in Antarctica, and he has been particularly concerned to check the ice-surge theory. During an expedition to McMurdo Sound on the Ross Sea in the southern summer of 1973-4, Denton and his colleagues found evidence of an advance of ice in the Ross Sea. Judging by fossils scooped up by its motion, the advance occurred during the most recent ice age. So the advance was apparently a result of the ice age, not a cause of it.

The research of Denton and others in Antarctica raises, instead, a worry of the opposite kind. Since the start of the present generally warm period, ten thousand years ago, the West Antarctic ice sheet has shrunk. Because it is a hangover from the most recent ice age, its melting is scarcely affected by the lesser fluctuations in climate from one century to another. The process still goes on and, so far from building and threatening to surge, the ice of West Antarctica may be going to melt completely. That would raise the sea level all over the world by about 15 feet. Even if it continues very rapidly by ice-sheet standards, Denton thinks that the melting of West Antarctica will take several centuries. 'So the Dutch need not be building their Arks just yet?' I asked him. 'No,' he replied, 'but perhaps they should be thinking where the wood will come from.'

The present threat of higher sea levels could be overtaken by much cooler conditions that bottled up more ice in the world. And although Denton himself remains sceptical, the possibility of an ice surge in the Ross Sea during a warm period like the present cannot be completely ruled out. At the bottom of the West Antarctic ice sheet is a layer of water and, in some places at least, the ice is now growing thicker rather than thinner. As an explanation for the major ice ages, the ice-surge theory looks weak nowadays, but it still might account for the sudden cooling of 90,000 years ago. Willi Dansgaard himself doubts whether an ice-surge could take effect quickly enough on the climate of the northern hemisphere to cause changes as rapid as those he detects.

As Dansgaard was showing me the Greenland record of the sudden cooling which began in conditions not very different from our own, he remarked gravely: 'We must find the explanation.' But seen amid the succession of major ice ages through which the Earth is passing, the sudden-cooling event and the possible melting of West Antarctica are two minor phenomena. They have provided a convenient introduction to some of the methods and preoccupations of modern research on major climatic changes; also to the slow-motion behaviour of ice sheets, which contrasts so greatly with the continually changeable weather of the air. They reinforce, too, the discoveries that, in the past few years, have utterly transformed the estimations of the threat of a new ice age.

Frequent freezing

Advice to young geologists: go and see for yourself and on no account believe what your professors tell you. The disturbing facts about the ice ages have emerged only slowly, because of the conservatism of experts. This century has been a marvellous time for discoveries about the Earth, but a dreadful one for geologists. As a profession, and most vehemently in the United States, they resisted the evidence for continental drift as well as for frequent ice ages. They have had to eat a great many learned words in the past few years, and scrap a great many textbooks. It fell to an outsider, the German meteorologist Alfred Wegener, to amass the arguments for continental drift, and to bear the scorn of the outraged professionals. Similarly a young Italian-born hunter of marine fossils, Cesare Emiliani, produced the first clear evidence for a rapid and strikingly rhythmic succession of ice ages. Unlike Wegener, who died on the Greenland ice sheet, Emiliani has had the satisfaction of seeing his discovery accepted, after suffering his share of scorn for nearly 20 years.

The first substantial clue that the ice was formerly far more extensive than at present was spotted two centuries ago in Europe: huge blocks of granite lying on

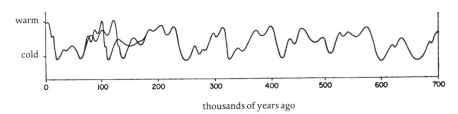

warm

cold

0   100   200   300   400   500   600   700

thousands of years ago

the slopes of the Jura Mountains had come from the Alps, 50 miles away. When Horace de Saussure of Geneva recognised them he said (conservatively) that they had been swept there by Noah's Flood. James Hutton, the founder of scientific geology, grasped the true meaning of these gigantic boulders. In 1795 he described the Alps as formerly covered by a mass of ice, with immense glaciers carrying blocks of granite for great distances. Nobody listened.

In 1837, the young naturalist Louis Agassiz said it again. Ice had plainly polished and streamlined rocks far below the existing glaciers, or deeply furrowed them with the boulders that it carried. Heaped rocks (moraines) marked the limits of former glaciers. Despite ridicule and even anger from eminent geologists, Agassiz persisted. On a visit to Scotland, which has no permanent snowcaps or glaciers nowadays, he found the same kinds of polished rocks and heaped moraines. They spoke of a huge ice field that had covered the high ground and extended down the valleys. Later, he went to North America and no sooner had he stepped ashore in Halifax, Nova Scotia, than he found the same telltale marks of a former ice sheet.

Gradually geologists accepted the ice theory; it explained many familiar features of the northern landscapes that had baffled them for years, notably the peculiar clay, stuffed with stones, that the ice smeared over vast areas of Europe and North America. In the mountains, characteristic handiwork of the ice includes jagged peaks and ridges and loose fragments (scree) formed when frost shatters the rocks; circular hollows (cirques) quarried by the ice, and deep U-shaped valleys carved by glaciers. Lakes formed by melting ice range in size from the 55,000 small lakes of Finland to the Great Lakes of North America and the Baltic Sea. When the North Sea rose as the ice melted, salt water spilled into the Baltic. The fjords of Norway, Scotland,

Greenland, Canada, Alaska, New Zealand and Patagonia are glacier valleys now flooded by the sea.

The shrewder converts quickly realised that the ice did not come and go in a single episode. Successive layers of the 'boulder clay', for example, separated by soil or peat, spoke of a succession of ice ages. (Some experts call them 'glaciations' and mean by 'ice age' the single continuous period covering the warm intervals like our own as well as the cold ones; I am using 'ice age' in its more popular sense, which is also accredited by other experts.)

In 1909, Albrecht Penck and Eduard Brückner marshalled the evidence of successive periods of growth of the Alpine glaciers. Their tome, *Die Alpen im Eiszeitalter*, affirmed that there were four ice ages, no more and no less. It took sixty years to break the spell cast by that book over the minds of geologists. The ice ages in the Alps took their names from Bavarian streams that exposed traces of particular episodes of glaciation. For the corresponding freeze-ups in North America and elsewhere other names were invented, for example:

Alps: Würm (youngest)	N. America: Wisconsin
Riss	Illinoian
Mindel	Kansan
Gunz	Nebraskan

All sorts of snags soon appeared in many countries, but the doctrine of four ice ages was so strong that contradictory evidence was explained away or ignored, sometimes even suppressed, rather than let it compromise the magic number four. At best, there was tolerance for the idea of more than one period of extreme cold within each ice age, and for hints of an older ice age before the Gunz.

In the early 1950s, Cesare Emiliani was working at the University of Chicago. There the celebrated chemist Harold Urey was exploiting, in all sorts of ways, the

zu den heutigen Umständen, sondern sie beschreibt Oszillationen. Es interferieren mit den grossen Schwankungen der Glacial- und Interglacialzeiten kleinere, über deren Umfang wir noch im Unsicheren sind. Wir erhalten jene Klimakurve (Fig. 136), indem wir als Abszissen die ungefähre Dauer der einzelnen Zeiträume des Eiszeit- alters auftragen; dabei setzen wir die uns noch unbekannte Dauer der Würm-Eis-

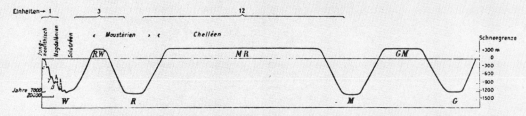

Fig. 136. Klimakurve des Eiszeitalters.

zeit provisorisch gleich der der Riss-Würm-Interglacialzeit, die der Riss- und der Mindel-Eiszeit aber etwas grösser an; wir nehmen ferner an, dass alle Übergangszeiten von gleicher Länge gewesen seien wie die Post-Würmzeit, an deren Ende wir leben und die uns als Masseinheit der Schätzung der Interglacialzeiten diente. Letztere sind bei dieser Schätzung von einer Eiszeit zur nächsten, also einschliesslich der Über- gangszeiten gerechnet worden. Als Ordinaten tragen wir die Schneegrenzhöhen ab, die wir für die einzelnen Eiszeiten und Rückzugsstadien ermittelten; für die Inter- glacialzeiten nehmen wir dabei allgemein den Wert, den uns die Höttinger Breccie gewährte, ohne damit behaupten zu wollen, dass alle Interglacialzeiten gleich warm gewesen seien. Für die Interstadialzeiten deuten wir durch doppelte Zeichnung der Kurve die Unsicherheit unseres Ergebnisses betreffend die Grösse des Rückzuges an.

improved techniques for weighing atoms. Chapter 1 described Willi Dansgaard's work in Denmark, using a 'mass spectrometer' to measure the changing propor- tion of heavy oxygen atoms in successive layers of the Greenland ice sheet. The method originated in Chicago, where it was applied not to ice but to the chalky fossils of animals that lived in the ocean and sank to the bottom when they died.

By a lucky coincidence, Swedish and American oceanographers had begun collecting many samples of mud from the ocean bed. They employed a 'piston corer', a tube that drove itself into the soft ocean bed and took up a core of mud many feet long. In a core brought from the bottom of the Caribbean, Emiliani picked out fossils of small animals that he knew had lived near the surface. He then measured the propor- tion of heavy oxygen in a mass of the little fossils. He worked his way through the different layers of mud – the deeper the layer, the older the mud and its charge of

fossils. By Emiliani's interpretation a high proportion of heavy oxygen is evidence that the animals lived in cool water, and vice versa (the opposite of the relation- ship found in the Greenland ice).

The Caribbean had gone through at least seven great cycles of cooling and rewarming. Of course, there were no great ice sheets in the Caribbean area, even in the depths of the ice ages, but Emiliani realised that the repeated rises in heavy oxygen in his fossils corres- ponded to periods of general cooling of the Earth and the spread of the ice sheets over the northern lands. Emiliani announced this discovery in 1955. But the number seven was obviously too great, by the conven- tional ideas about the ice ages.

The fossils in other ocean-bed cores showed just the same patterns of cooling and warming, and aroused great interest among the non-experts. The variations in temperature were amply confirmed because the animals themselves were alternately of species that liked warm

water and of other species that liked cool water. Emiliani's case was not helped by the fact, as we now know, that he was greatly underestimating the ages of his older specimens. That did not alter the evidence for 'more than four' ice ages, but the specialists were unmoved. As recently as 1968 some researchers were still busy trying to force the ocean fossils to fit the theory of four ice ages.

What were a few small sea animals, compared with the Alpine moraines, the boulder clays and the other features delineated by Penck and Brückner in 1200 pages of German text? With hindsight we can say that the areas affected by the ice sheets were the worst possible places to look for the record of climatic changes during the ice ages – for the simple reason that each new ice sheet or advancing glacier thoroughly mangles the traces of its predecessors.

The evidence on land that was to vindicate Emiliani turned up in Czechoslovakia. Although hemmed between the ice sheets of the Alps and of northern Europe, Czechoslovakia escaped burial under the ice. Instead, in the depths of an ice age, the country was a frigid desert like the Gobi Desert today. Across it, as in the Gobi, the wind blew clouds of fine dust, which covered the land with a yellow layer of 'rock flour' or loess. When the climate improved again new, dark-coloured forest soil formed on top of the loess. There were other phases that left relics of grassland soils and deeply frozen ground. But the main alternations between very cold and dry and very warm and moist conditions, in the soil-loess-soil succession, happened more often than four times.

Where rivers cut through the layers, or humans dig for the loess to make bricks, they expose the tiger-pattern of yellow and darker layers in the ground of Czechoslovakia. In these stripes in 1960 George Kukla from the Geological Institute in Prague began taking

stock of the evidence for many ice ages. At more than 30 sites, clustered around Prague and between Brno and the Austrian border, Kukla and his Czechoslovak colleagues studied the layers and the traces of life they contained. Different species of snails kept changing the guard as the cold came and went; fossil pollen showed rich forests alternating with chilly grassland; and woolly rhinoceroses lived at Prague during the last ice age. Apart from the signs of at least ten complete cycles of cooling and rewarming, the most striking feature was the steady rhythm of the changes.

Dating was the difficulty. In the Czechoslovak soils, as in the ocean-bed cores, the scientists tried several techniques for assigning ages to the various layers, but they were insecure if not positively misleading. There was no correspondence – except in their obvious multiplicity – between Kukla's cycles and Emiliani's. Help eventually came from the skittish behaviour of the Earth's magnetism.

Every so often the Earth goes through a great flip, swapping its north and south magnetic poles around. The layers that slowly accumulate on the ocean bed and on land are very weakly magnetised by the prevailing magnetic field of the Earth and they preserve a record of it. One magnetic flip, well known to earth scientists, occurred 700,000 years ago. Before then, the magnetism had been opposite to what it is today; afterwards it was the same as today. The reversal was ideally timed, in the midst of the series of ice ages, to give the dating schemes a trustworthy anchor.

In 1969 the magnetic reversal was reported in the natural climatic records both of the ocean-bed cores and of the Czechoslovak soils. At the Lamont-Doherty Geological Observatory near New York, sensitive magnetic measurements found the reversal in four cores from the equatorial Pacific. The varying abundance of chalky fossils in the cores spoke of eight great ice-age

In the ground of Czechoslovakia, light-coloured soil (loess) corresponds to a time when the region was a cold, dust-blown desert. To George Kukla (right) such layers on land confirmed the earlier evidence from the ocean bed – that ice ages were much more frequent than had been supposed.

cycles during the 700,000 years since the reversal. In the same year, the Czechoslovak scientists reported the magnetic reversal among their soil layers. Counting down from the present, it was just below the eighth major period of cooling. Bingo! The evidence of drastic changes of climate by land and sea, which had been in conflict for years, now matched.

While this was going on, Czechoslovakia was passing through the Prague Spring of the communist reformation and then the Soviet invasion followed by home-grown repression. Kukla is now a political refugee at the Lamont Observatory near New York, which was a former stronghold of the four-ice-age theory. His presence there has done more than any amount of publication of evidence to change the minds of experts.

Two hundred years after the discovery of those misplaced boulders from the Alps, the story is plain at last. Some lesser difficulties persist with the dating of some of the events since the magnetic reversal of 700,000 years ago, but the main conclusions can no longer be contested. There have been eight major ice ages in that period and perhaps a dozen previously. Unlike many squabbles among experts the question of the number of ice ages is a matter of practical importance and, although a triumph for science, the answer is ominous for the human species.

According to the old Würm-Riss-Mindel-Gunz theory, each of the four ice ages lasted about 100,000 years while the intervening warm periods or 'interglacials' endured for 125,000 to 275,000 years. As the present warm period started as little as 10,000 years ago, mankind could look forward to a generally warm climate for more than 100,000 years into the future.

The new picture is completely different. Between the rapid succession of icy periods, the really warm periods such as the one we live in have lasted about 10,000 years at the most. That puts a changed complex-

ion on the fact that the present warm period began *as much as* 10,000 years ago. And I shall have to report, later on, another discovery in the ocean-bed cores that further alters our human expectations for the worse. Meanwhile, we should look more closely at what ice ages are like, and why they happen.

Portrait of an ice age

Only Antarctica and Greenland carry massive ice sheets at present, and Antarctica's are roughly ten times as big as Greenland's. All the other glaciers in the world – and there are many of them – represent only a hundredth of the semi-permanent weight of ice on land. Long ago, before the ice overwhelmed them, Antarctica and Greenland were continents like any others, possessing hills and valleys, plants and animals. Today one can only grope for those lost landscapes. Special airborne radar, for example, penetrates the ice and recovers echoes from the underlying rocks. Echoes also come from mysterious layers within the ice which stretch for hundreds of miles and seem to signal some change in the weather at particular times in the past – although no one is sure about what kind of change it was.

During an ice age, the Antarctic and Greenland ice sheets grow proportionately very little. Scarcely any new territory remains for the ice to occupy before it runs into deep ocean water and becomes icebergs. But huge new ice sheets form, on northern lands, trebling the total area of the world's ice sheets. Southern Alaska, virtually all of Canada and the northern fringe of the United States disappear under one ice sheet, in area as large as Antarctica, and typically 3000–4000 feet thick. During the most recent ice age, a salient of this ice sheet stretching past Chicago and Detroit reached as far south as Columbus, Ohio; on an earlier occasion a similar salient reached St Louis, Missouri. In the west, it

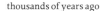

thousands of years ago

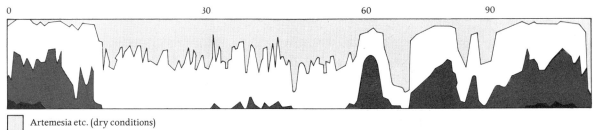

0   30   60   90

☐ Artemesia etc. (dry conditions)

■ Oak pollen (wet conditions)

merged with the ice caps on the Rockies and other mountains, which extended in a long succession of glaciers to Mexico City.

The European ice sheet stretches from Scandinavia south-west across the North Sea to Britain and east into north-west Russia. In the most recent ice age it reached to Hamburg and Berlin in the south and the Urals in the east, but in previous episodes it reached Leipzig and joined up with other ice far across northern Russia, also making a great extension down the Dnieper to within 200 miles of the Black Sea. The Alps, from the Rhône above Lyon in France to Graz in Austria, become a big lump of ice. The ice sheets in Siberia and the Himalayas are smaller than might be expected. So are those of the southern hemisphere – they contribute only 3 per cent of the world's area of new ice, during an ice age, because there is so little land at the right distance from the South Pole.

But an ice age is not just a load of new ice heaped on the northern lands. Some parts of the world benefit. With water locked up in ice on land, the sea level falls, sometimes by as much as 600 feet. The territory of south-east Asia, from China to Java, was wonderfully enlarged and the spit of drowned land between Australia and New Guinea broke surface too. The whole weather machine alters, and in ways only roughly understood so far. So what was the world like, only 18,000 years ago, when it was deep in the most recent of the ice ages?

The equatorial zone of the Earth had turned dry. An old idea was that whenever the world was icy in the north, it was rainy in the tropics and sub-tropics, with the so-called 'pluvial' and 'glacial' phases neatly matching. Sometimes there was indeed a great deal of rain, and lakes like Chad in Africa and the Caspian Sea in Russia swelled prodigiously at times. But more detailed histories of lakes in equatorial Africa (Lake Victoria) and Colombia (Laguna de Fuquene) have become available in the past few years. They show, during the last phases of the most recent ice age, long periods of drought alternating with shorter periods of very heavy rainfall. One period of equatorial drought starting 20,000 years ago lasted 7000 years. The Fuquene region of Colombia in South America, now cloaked in mountain forest, was then chilly, dry grassland.

For an impression of what the Mediterranean's climate was like when northern Europe was in the grip of the ice, we can look to a deep boring made in marshy land in Macedonia in northern Greece. Since the end of the most recent ice age this region, according to the pollen record, was naturally an area of oak forest. But 18,000 years ago virtually no trees were present and what there were were pine. The climate was dry as well as cold, although not far away the Jordan Valley was flooded at that time.

A group of pollen hunters from the University of Amsterdam, led by Thomas van der Hammen, has done the work just mentioned, in Colombia and Greece. They have also carried out widespread studies in their own country. Often they bore deep holes to recover the fossil soil and pollen from past climatic periods, but sometimes rivers have done the digging for them. The Dinkel river, near the German border, makes a unique geological monument. It cuts through overlying sand and displays, for all to see, a relic of the severe phase of the most recent ice age. A polar desert, as on Arctic islands today, is preserved as a conspicuous band of loamy sand, three feet thick, with big cracks due to freezing.

The pollen record of Europe, back through the warm and cold periods of the ice-age cycles, shows an amazing alternation. In the warm periods the continent has been cloaked in forest – primarily oak in the south and conifers and birch in the north – right up to the Arctic Circle. But when the ice sheet covers the northern part

Greece was dry during the last ice age, as shown by a diminution of oak pollen (after T. A. Wijmstra, University of Amsterdam). Below are some tree and shrub pollen grains found by the Dutch scientists in Colombia, South America; also a view of present-day forest, 9000 feet up in the Andes, which gave way to open vegetation during the ice age (compare with the colour plate facing page 96).

of the continent and the Alpine glaciers are licking the surrounding landscape with their icy tongues, much of the continent is without trees. In southern France, instead of vines, Lapland-like birches grow.

Not St Patrick but the ice freed Ireland from snakes. Living species in Europe have had an exceptionally hard time during the ice ages because of the barriers of mountains and the Mediterranean Sea that ran east-west across the continent. Many a species of plant or animal has shifted its ground southwards as the climate has deteriorated, keeping ahead of the frost as it were, only to be brought up short by frozen ranges of mountains – if not by the Alps then by the Pyrenees or the Carpathians or the Caucasian Mountains – or else by the sea. Species of plants that survived in North America and Asia, with southerly 'escape routes', were annihilated in Europe. Yet the isolation of small populations of plants and animals, leading difficult lives in pockets of refuge, was a powerful stimulus to the origin of new species peculiar to Europe. Some, including the great cave bear and Neanderthal man, appeared briefly, only to be extinguished again. Modern men, like ourselves, advancing into Europe from their origin in the Middle East 50,000 years ago, have been blamed for their demise, but the climatic changes of the time were acting more ferociously than small tribes of humans ever could have done. The strip of polar desert in the United States was only about 100 miles wide, between the ice sheet and the chilly spruce forest that covered a large part of the country. On the west coast of North America, temperatures were about 6 degrees lower than today, but that was much less of a fall than on the west coast of Europe. One reason is that those regions are already cooler, because they lack the Gulf Stream; another is that, soon after the ice age started, the Bering Strait between Siberia and Alaska dried out and cut off the Pacific from the Arctic Ocean.

The vigour of the trade winds during the ice ages can be judged by their success in transporting dust from West Africa over the Atlantic. The stronger the wind, the coarser are the particles that are carried, say, 500 miles out to sea. In the ocean bed off Africa, relatively coarse particles are more abundant during the ice ages. A simple argument says that the trade winds are stronger at times of world-wide cooling, because there is a sharper temperature contrast between the tropics and the adjoining regions nearer the poles. But there are puzzles. One is that vigorous trade winds are hard to reconcile with the ice-age droughts in equatorial regions – because the cloud clusters that pump air upwards and suck in the trade winds are also the source of heavy rain.

Another puzzle is that vigour in the global winds is thought to mean efficiency in transporting heat from the tropics towards the poles. A vigorous main jet stream and strong eddies in the stormy zone are a warming influence on the northern lands. A weak jet stream characterised the Little Ice Age of recent centuries. And some experts blame the dryness of Europe, which made all that dust during the real ice ages, on a prevailing north-east wind – a block, as it were, that lasted thousands of years. More careful comparisons of evidence from different times and places are needed, together with a shrewd use of computers, to reproduce the ice-age winds. Any real grasp of how and why ice ages differ from warm periods like the present, or how and why ice ages begin and end, will come only with a much clearer picture of the variable patterns of the global winds, of temperatures and of snow.

The most ambitious study of ice ages now in progress aims at just such an overall portrait of the world. CLIMAP is a joint project of four American universities – Brown, Columbia (Lamont Observatory), Oregon State and Maine. It uses the records of the ocean bed to

map climatic conditions and current systems throughout the world's oceans at chosen 'moments' among the past fluctuations. When combined with corresponding information from the land – about the extent of the ice sheets, for example – these maps will enable meteorologists to attempt to compute the weather throughout the world. They can then check on whether the ice-age weather so computed is correct, according to the knowledge gained from fossil pollen and the like. The aim is a much deeper insight into the workings of the weather machine in different modes.

The first 'moment' chosen is 18,000 years ago, when the most recent ice age was at its fiercest. By 1974 the work on this period had reached the point where the first computer runs could be attempted, reproducing within the limits of the numerical models the global climate of the time. Other 'moments' for the CLIMAP project are in the warmest phase since the most recent ice age (6000 years ago), in the warm period before the most recent ice age (120,000 years ago), and a much earlier time when the Earth was about to enter an ice age and for which a reversal in the Earth's magnetism gives a very reliable time-marker (700,000 years ago).

Great archives of ocean-bed cores – thousands of them – gathered over the years by ships of the Lamont Observatory and Oregon State University provide most of the raw material for CLIMAP. The ships are still hard at work, recovering a new core every day. From cores old and new hundreds of good ones are picked out, which cover the world's oceans. In each core, the mud of the 'moment' – 18,000 years ago in the first exercise – has to be found. A combination of dating procedures makes use of radioactivity, evolutionary changes in the fossil animals, and volcanic ash and magnetic reversals of known ages. The mud and its contents undergo all sorts of examinations and chemical tests, and measurements of the varying rates of accumulation of the mud.

But the main objective is to find the pattern of water temperatures.

Taking the temperature of the sea surface at a given spot, 18,000 years ago, is essentially a matter of seeing what kinds of animals and plants lived there. Just as you might judge the climate of a place according to whether polar bears or camels frequented it, so the appearances and disappearances and relative abundances of small marine species that lived near the sea surface tell how warm or cold it was. To adapt this general idea into a precise thermometer the fossil-hunters need a thorough knowledge of the present-day ensembles of species that inhabit water at different temperatures, and of how the individual species have evolved through time. Then very careful statistical sampling of the different kinds of small fossils in a given lump of mud is necessary before the temperature can be estimated. Mathematical theories describe the responses of the populations of different species, not just to prevailing conditions but to a history of climatic change. Valuable fossils in this work include foraminifera, small animals with chalky shells that abound in the world's oceans. One problem is that, if they sink into very deep water, chalky shells tend to dissolve.

From methodical work of this kind have emerged world maps of the sea-surface temperatures of 18,000 years ago. Subtleties in the changes they disclose at once demonstrate their value. In the tropical Atlantic off Sierra Leone, West Africa, for example, the temperature was 19 degrees instead of 26 degrees in winter today, but on the other side of the ocean, off Barbados, the sea was only two degrees cooler than today. The CLIMAP maps are a marked improvement on older versions of the ice-age history of the oceans which merely assumed that temperatures fell by the same amount all over huge expanses of water.

The middle zones of the oceans, between the tropics

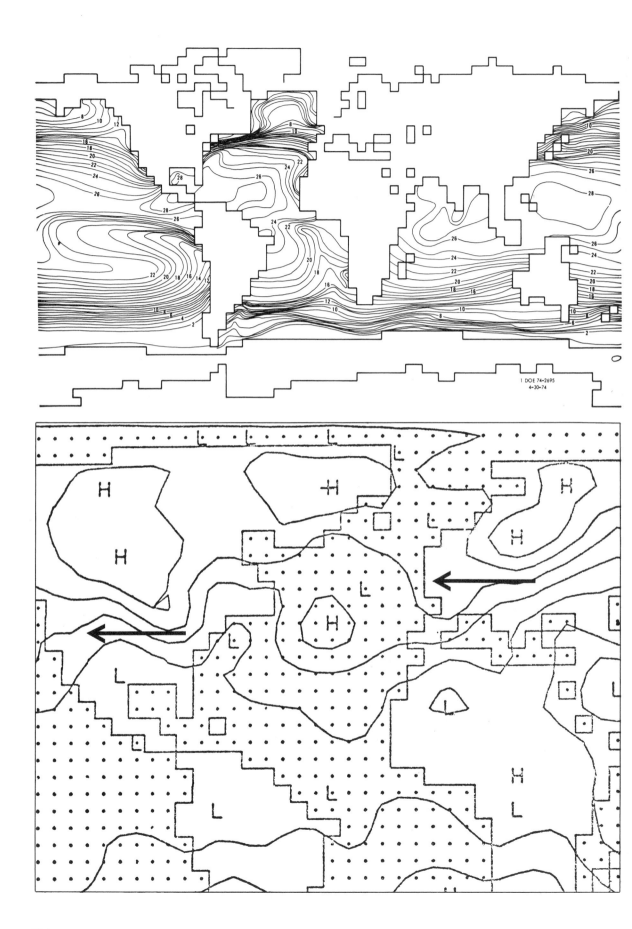

1 DOE 74-2695
4-30-74

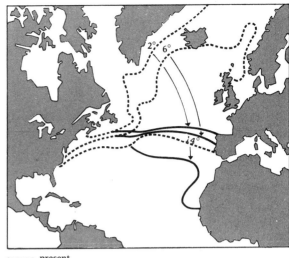

The CLIMAP project. Sea temperatures deduced for the worst of the last ice age (top left) have been fed into a computer that uses the Minz-Arakawa model to calculate the weather 18,000 years ago. Part of the result for typical surface pressures in July (bottom left) shows the presence over North America and Europe of strong easterly winds. They imply a cold, dry climate in both continents.

------ present
—— Ice age

Europe left out in the cold. The southward shift of cold Arctic water 18,000 years ago implies a much more southerly track for the warm water coming from America in the Gulf Stream.

and the polar chill, are the regions of the most revealing changes in sea temperatures, from the warm to the icy periods. And the most significant sector is the North Atlantic. There the change was more profound than just a general decline in temperature by several degrees. In the North Atlantic there was a qualitative change – a drastic alteration in the behaviour of the currents. The warm water of the Gulf Stream, instead of heading north-east from Florida to the British Isles and Scandinavia, drifted weakly east towards Morocco. As a result, the sea off northern Spain was as cold (2 degrees) as it is close to the Greenland sea ice today. A slightly southerly track for the Gulf Stream occurred during the Little Ice Age of recent centuries, but that involved a one degree fall in the sea temperature off the north-west corner of Spain. In the real ice age the fall was 10 degrees. The door of the Gulf Stream slammed shut and left Europe out in the cold.

The sea-temperature changes off the eastern United States were far less dramatic. In the ice age, as now, the warm Gulf Stream met the cold Labrador current coming south. The sea temperature off Cape Hatteras (North Carolina) was no colder than it is in winter today, despite the presence of the ice sheet only a few hundred miles to the north over New York and New Jersey.

New assessment of the ice of the period, together with the sea temperatures, went to Lawrence Gates at the RAND Corporation in California. He used a numerical model of climate to reckon the global weather in the ice age. The initial computations (March 1974) produced strong wings blowing from the east across North America in summer. That contrasted with the southerly winds of present summers. In Europe, too, there were strong easterly winds. For both continents that meant a dry climate as well as a cold one.

These reconstructions of world-wide patterns of weather in the ice age are only a beginning. One 'mo-

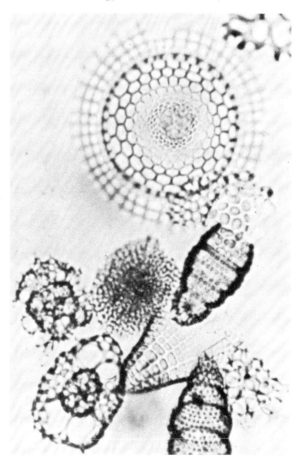

Some small fossils of the marine organisms that sank to the ocean bed when they died. Their changing fortunes make possible the reconstruction of past climates in the sea, as in the CLIMAP project.

Stones for Stonehenge (right). The arrows show ice flowing over the Bristol Channel (located left) and the broken lines are routes for transported rock.

Below right: The English Channel glacier, at the south-western edge of the European ice. It existed about 140,000 years ago, flowing from snow heaped on the dried-out sea bed south of Ireland. (Maps after G. A. Kellaway.)

ment', even if it were understood perfectly, would not be enough, because the climate was very different in successive phases of the most recent ice age. During the advance, remission, readvance and eventual retreat of the ice the patterns of rain and snow, and by implication the vigour and course of global winds and currents, varied greatly. The chill of an ice age suppresses much of the snow that nurtured it. Each ice age seems to have followed its own idiosyncratic course, in its timetable, its remissions and the extent of the ice sheets. There will be no quick or easy answer to the simple question: how did the weather machine work in the ice ages?

Ice from the sky

Concerning the ice that assailed the British Isles, old ideas have fallen apart, to be replaced by bolder new ones. This is not a parochial matter, because it helps to revise basic notions about where and how ice sheets form and spread. Conventional thinking about the ice ages in Britain had the ice starting in the northern mountains and spreading south. In the most recent ice age the ice was in any case largely confined to Scotland, but in the one before that (150,000 years ago) the ice reached Finchley just north of London, as well as burying Wales and most of Ireland.

Around 1970, these ideas began to quaver a little. The standing stones of ancient Stonehenge in southern England seemed perhaps to have been transported from the mountains of south-west Wales not (as archaeologists reasoned) by an elaborate stone-age transport system, but by an ice sheet. The English ice sheet had been more extensive than supposed. On a broader scale, European geologists agreed that the British and Scandinavian ice sheets were not separate entities, but were parts of a single ice sheet covering the North Sea – not

floating ice, but ice standing on the sea bed. As the first ice sheets formed and the sea level dropped, the shallow seas of the continental margin became available as new lands. In the snowy zones fresh ice sheets could grow and move on them.

In 1969, excavations for a road cutting in Somerset produced huge quantities of clay and stones dumped there by an ice sheet, 350 feet above the present sea level. It might have been just a striking extension of the previously known English ice sheets but for one remarkable fact. The flow of ice came from the *west*. For Geoffrey Kellaway of the Institute of Geological Sciences in London this discovery set off an exhausting train of thought and investigation in which his discomfiting of the archaeologists about the origins of Stonehenge was only a beginning.

By 1974, he was able to report that the English Channel was a former glacier. It was a proposition as revolutionary for the coastal geology of England and France as for ice-age theory. Yet in one swoop it cleared up dozens of mysteries. The evidence had been staring geologists in the face since Agassiz's time, including the erratic boulders littering the English shoreline (the sea-bed too) and the glacier clay smeared even on the sun-kissed Scilly Isles. The most extraordinary inversion concerns a layer of stones and sand at Goodwood in Sussex, 130 feet above the present sea level. Geologists had recorded this as a beach, deducing that it represented an exceptionally warm period between the ice ages when much of the Antarctic ice had melted, raising the sea by that amount. It was formerly one of the anchor points of world-wide ice-age studies. Now Kellaway recognises the layer as a mark of an exceptionally *cold* period: not a beach, but debris washed out of the flank of the English Channel glacier.

But Kellaway also declares that the ice did not come off the English hills, or squeezing through the narrow

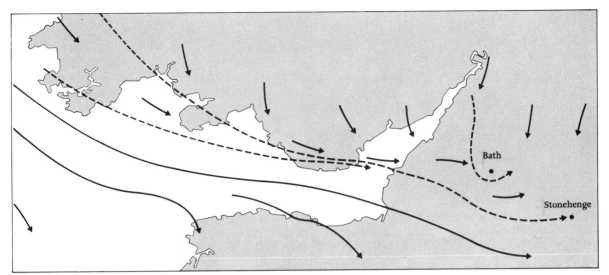

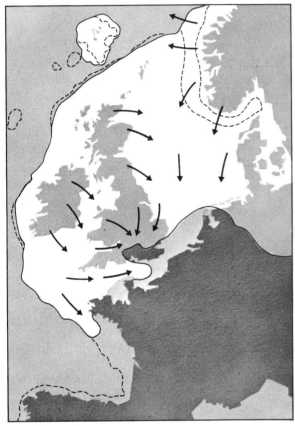

— ice margin
---- continental shelf

Dover Strait from the North Sea, but quite the other way. It came from what is now the shallow bed of the Western Approaches beyond England's south-west tip. That west-to-east flow of ice over Somerset was just a sideshow in a much greater kind of event. The ice spread up the dry bed of the English Channel at least twice and most recently around 150,000 years ago, during the last ice age but one.

For achieving this huge addition to the ice of Europe the weather machine's trick is as follows. First form some ice sheets on land, thereby bottling up water and lowering the sea level. As the Western Approaches break surface, pour on to the new land abundant snow, made from the water vapour of the Gulf Stream – which, although deflected well to the south, is only a short wind-ride away. A new ice sheet forms, accelerating the fall in the sea level. It may calve icebergs into the deep Atlantic water south of Ireland, but some of the ice will spread eastwards, up the English Channel at least as far as Sussex, adding a long list of south-coast resorts to the places that were once buried under ice. Kellaway envisages two other great ice domes on the dried sea bed: one north of Ireland and the other between Scotland and Norway.

Now is the moment to introduce two contrasting theories about how ice sheets form and grow in an ice age. The older one was born of glacier-watching and the belief that ice sheets start to form on high ground and spread outwards. I call this theory the snowtrap. Ice sheets are made of snow and everyone sees snow forming and lying as snowcaps on mountains. Snow that falls on top of previous snow or ice has less chance of melting than does snow falling on uncovered ground. So the picture is of a small, embryonic ice sheet acting as a snowtrap and thereby growing thicker. As it does so, it creeps outwards; that enlarges the snowtrap, which captures more snow – and so on. In short, the snowtrap theory has an ice sheet spreading gradually outwards from its core, which is generally presumed to be a range of mountains.

There is no doubt that ice sheets do tend to spread sideways as their thickest parts sag under their own

Left: the greatest extent of the northern ice sheets – so far – during the present series of ice ages. (Adapted from a map by R. F. Flint.)

Right: eight small shells of animals that lived on the ocean bottom, here separated from the others that lived at the surface and sank when they died, are enough for a measurement of the volume of ice in the world when they were alive.

weight. That is how they smear the land with clay and boulders carried from far away, and push glaciers down convenient valleys. To that extent the snowtrap theory is at least partly correct. But as described it is a very slow process, taking 15 to 30,000 years to build a major ice sheet across a continent. In human terms, you would see the ice sheet inching towards you on a distant skyline and warn your grandchildren that they might have to think about moving a few miles further to the south.

The second theory is gathering strength. I call it the snowblitz – meaning 'blitz' in the sense of *Blitzkrieg*, or lightning war; nothing to do with lightning as such. In the snowblitz the ice sheet comes out of the sky and grows, not sideways, but from the bottom upwards. Like airborne troops, invading snowflakes seize whole counties in a single winter. The fact that they have come to stay does not become apparent, though, until the following summer. Then the snow that piled up on the meadows fails to melt completely. Instead it lies through the summer and autumn, reflecting the sunshine. It chills the air and guarantees more snow next winter. Thereafter, as fast as the snow can fall, the ice sheet gradually grows thicker over a huge area. In human terms, your land is snatched from you in a single bad summer and there is nothing to do but trek across the unmelting snow in search of fertile territory. It might be many miles away and already swept by chilly winds off the wafer-thin ice sheet.

The snowblitz is thus a very much faster and more alarming way of establishing an ice sheet than by the outward creep of a snowtrap. The snowblitz theory was advanced in 1970 by Hubert Lamb and Alastair Woodroffe, in the course of an investigation into the global-wind system during ice ages.

One way of trying to see which theory is more likely to be right is to find out how quickly the ice sheets have grown in the past. At present the Earth's ice sheets are equivalent to a giant's ice-cube measuring about 200 miles on each side. In the depths of the ice age they are nearly three times as large, by volume.

Nicholas Shackleton of the University of Cambridge can estimate the size of the ice sheets, at a given time in the past, from a set of just eight small shells, each four thousandths of an inch across, the remains of sea animals that lived far away from any ice sheets. Shackleton came into this work only because the strong Cambridge group of ice-age researchers solicited his skills as a geophysicist. But he has won a high reputation, not least for his ability to weigh the world's ice.

Behind this scientific sleight-of-hand is a simple idea: when ice sheets grow, the proportion of heavy oxygen in the water of the ocean increases. I have made much of the variations in heavy oxygen in the Greenland ice and marine fossils, as indicators of climatic change. But *all* the ice in the polar sheets is deficient in heavy oxygen compared with sea water. Water molecules containing heavy oxygen evaporate more sluggishly than ordinary water does and so are less likely to find their way into the snow that builds the ice sheets. The ice sheets represent a sufficient fraction of the world's water to alter by easily measurable amounts (around a tenth of one per cent) the proportion of heavy oxygen remaining in the sea. Indeed, that turns out to be the chief reason why the heavy oxygen varies in the marine fossils of the tropics, during the ice ages.

A complication is that the sea animals used in the measurements can also vary their uptake of heavy oxygen because the local temperature of the sea changes. Such was the assumption in Cesare Emiliani's pioneering work. To avoid confusion and find the changes due solely to the changes in the composition of the water, Shackleton picks out from the graveyard of the ocean-bed core those animals that lived on the

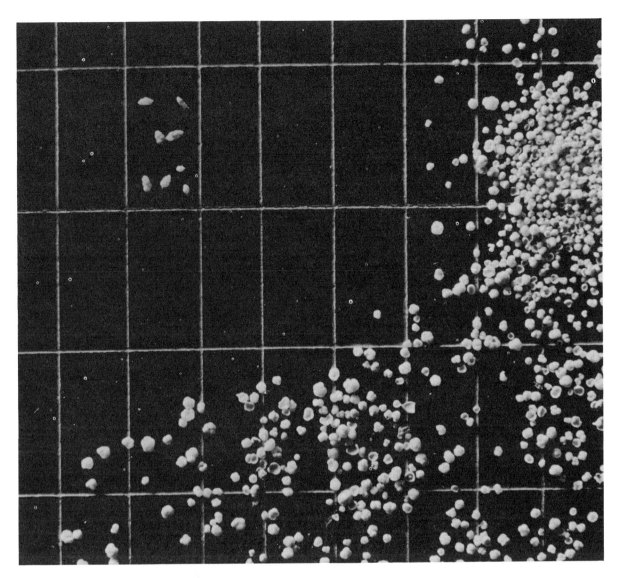

ocean bed itself. In the endless darkness under three miles of water, the temperature scarcely changes, whatever is going on at the surface.

The bottom-dwellers are outnumbered by the other animals that lived at the surface, and sank when they died. Shackleton has so sensitive a 'mass spectrometer' for weighing oxygen that he needs very few of his chosen species to make a measurement. The work of filtering the mud and sorting out, under a microscope, the bottom-dwellers from the rest is nevertheless finicky. It takes a quarter of an hour for Shackleton to find a quorum of eight or ten of them. But when the shells have been dissolved in acid and given off their carbon dioxide, and when the magnets have sorted the molecules containing heavy oxygen, the numbers coming out of the machine at Cambridge tell of the great ice sheets that altered the northern world beyond

recognition and sometimes buried Cambridge itself.

The general story of the ice sheets, as measured from the animals of the ocean bottom, corresponds very well with the ice-age records of the sea surface and the Czechoslovak soils. The eight ice-age cycles in 700,000 years, for example, are completely plain. But here Shackleton is seeing the waxing and waning of the ice sheets and what impresses him is the speed with which they grow.

The very slowly accumulating layers on the ocean bed give, as noted earlier, only a smudged record. They must tend to understate the rate and extent of the changes in the ice. In 1974, working with an American core from the floor of the Pacific off Ecuador, Shackleton carefully traced the end of a very warm period, about 110,000 years ago, as registered by his bottom-dwellers. He found that the world's ice sheets built up to fully

Helped by technologies like the ice-proof boat, Eskimos have adapted to the cold life, even though the ice stretches as far as the eye can see. The view from Atlantic City would have been like this in an ice age.

half their ice-age volume in about 5000 years.

That is far too fast for the snowtrap theory of gradually widening ice sheets. It presents no difficulty to the snowblitz theory, which has the ice coming out of the sky and building up ice sheets over a huge area from the outset. On this view the figure of 5000 years does not mean that the climate takes that long to deteriorate. The cold comes instantly, the snowblitz seizes huge tracts of land instantly, but then the snow piles up for 5000 years at perhaps 18 inches a year. 'Instantly' may mean a hundred years, or a single bad summer.

So ice ages can start very suddenly – that is the implication of this research and of the snowblitz theory. It contrasts sharply with the conventional idea – of which the snowtrap is the latest version – that the world's climate only slowly worsens as the ice spreads. There is other evidence. For example, 70,000 years ago, as the Dutch pollen hunters record, most of the trees of Macedonia and essentially all of the trees of the Netherlands died within a thousand years. Again, the record is smudged and the time could be shorter. The sudden cooling – an ice age that failed to stick – tells a similar story and then, according to Willi Dansgaard, the plunge into full ice-age conditions took no more than a hundred years.

The discovery of the English Channel glacier also gives powerful support to the snowblitz theory – because in this case a fresh pile of ice starts on the very lowest land, newly laid bare by the receding sea. The conventional ideas, born of glacier-watching, not only have the ice sheets starting from snow on the cold mountain-tops, but also spreading essentially downhill. Of course glaciers, like water, can only flow in such a way as to lower the top of the ice. But by the snowblitz theory ice can pile up on low ground where the snowfall happens to be heavier, and then flow downwards on to higher ground.

The trend towards a catastrophic view of the onset of ice ages makes a very big difference to any estimate of the possibility of a drastic change of climate occurring in the near future. But before considering what the risks may be, we should see how ideas stand today about the causes of ice ages.

The best of all possible worlds

A noted Soviet climatologist estimates that, if the ice sheets reached a little closer to the equator than they have managed so far, the ice age would become unstoppable. The entire Earth would become bandaged in ice. At a temperature below − 50 degrees, life would cease and the Earth might stay that way for aeons. By comparison, the ice ages and the warm periods like the one we live in are both conditions of the Earth that can be described as 'moderately icy'. In one case there is twice as much ice on land as in the other, but in both situations the tropics are warm, the poles are icy and substantial areas of fertile land and sea remain. In this perspective, the changes in the ice-age cycle are merely variations on an icy theme.

Just how fortunate in its climate the Earth is can be seen in a glance at our neighbour planets, Mars and Venus. The average surface temperature on Mars is 40 degrees below freezing, on Venus more than 400 degrees above the boiling point of water. Mars spins at about the same rate as the Earth but it has a rarefied atmosphere of carbon dioxide and possesses very little water. The effects of tides in the atmosphere are much more pronounced than on Earth. Extraordinary storms occur on Mars, which obscure the whole planet in dust for many weeks; they are similar to hurricanes but are powered by the sunlight absorbed by the dust, instead of by the heat released when water vapour condenses.

Venus has a very dense atmosphere of carbon dioxide and is permanently obscured by thick clouds; they are thought to be made of sulphuric acid. The atmospheric pressure at the surface is 90 times that on Earth. Venus rotates slowly, which means that its global winds are quite unlike the Earth's. Television pictures from *Mariner 10*, which passed close to Venus in 1974, show a 'hot spot' directly under the Sun and strong winds spiralling in towards the pole in each hemisphere. The heat catastrophe that has overtaken Venus seems to be due to the carbon dioxide and may have a moral for the Earth. If the Earth became substantially warmer − perhaps because of man-made carbon dioxide − the hot seas would release dissolved carbon dioxide, making the planet warmer still and releasing more carbon dioxide − and so on. In a vivid mixed metaphor this possibility is now being called the 'runaway greenhouse'.

There would, of course, be no ice, but an ice-free situation does not require so drastic a change. A complete melting of the present ice could occur without much increase in the overall warmth of the world − indeed that is one reason for concern about the present moderate increases in carbon dioxide in the atmosphere and the potential effects of much larger human use of industrial energy. West Antarctica, as mentioned, could by melting raise the oceans by 15 feet; if East Antarctica and Greenland are added, the figure becomes 200 feet and many of the world's most valued valleys and coastal plains, and all the seaports, would vanish beneath the waves.

So we can pick out, from all this, a remarkable range of possible conditions:

Earth 1. Superheated, like the planet Venus.
Earth 2. Ice-free, with a high sea level.
Earth 3. Slightly icy, like today's conditions.
Earth 4. More icy, as in an ice age.
Earth 5. Superfrozen, as a ball of ice.

During the ice ages, Colorado looked like this. The photograph shows Antarctic mountains standing shoulder-high in ice. These mountains divide the huge ice sheet of East Antarctica from the more changeable West Antarctic ice sheet.

This scene could be Iowa in an ice age. It is the present-day tundra of northern Russia, with dwarf vegetation braving the cold. The reindeer, at least, are able to migrate south for the winter.

There is reassurance in the fact that, despite enormous changes in geography and in the composition of the air, our planet has so far avoided the two extreme conditions and has merely varied between Earth 2, Earth 3 and Earth 4. There seem to be powerful natural regulators at work which maintain the tropical temperatures, at least, within fairly narrow limits and keep the all-important composition of the sea water more or less constant. But humans will be well advised to learn about these global regulators and make sure that they do not unwittingly override them.

The present state of the Earth, No. 3, is for mankind the best of all possible worlds, if only for the tautological reason that it represents the conditions in which our civilisations have developed and spread. There is enough ice in the world to keep the sea level down but not so much as to rob us of the northern lands. Weatherwise, the global winds are vigorous and they carry moisture and winter warmth from the sea, deep into the continents. It seems likely, from historical studies and numerical models, that we have a lot to thank the jet stream for; either warmer or cooler conditions in the north seem liable to weaken it. The present climate is right for agriculture or for abundant natural growth in 40 per cent of all the world's territory; 20 per cent is too arid, 20 per cent is too mountainous and only 20 per cent is lost because it is too cold. By these various tests, especially by the aridity of the polar and tropical deserts during an ice age, there is no doubt that Earth 4 would be a tragedy. You could make out a case for Earth 2, though, and if in doubt it might be better to let our cities drown, to err on the side of melting, if that were the price for avoiding the next ice age.

Through much of its long history the Earth has been entirely free of ice sheets. The sea level was correspondingly high and conditions were probably balmy. But beware: many of the stories that geologists told

Ice-free world. The map shows how much of present-day Europe would be drowned if the polar ice sheets were to melt completely, so raising the sea level by 200 feet. It might be preferable to an ice age.

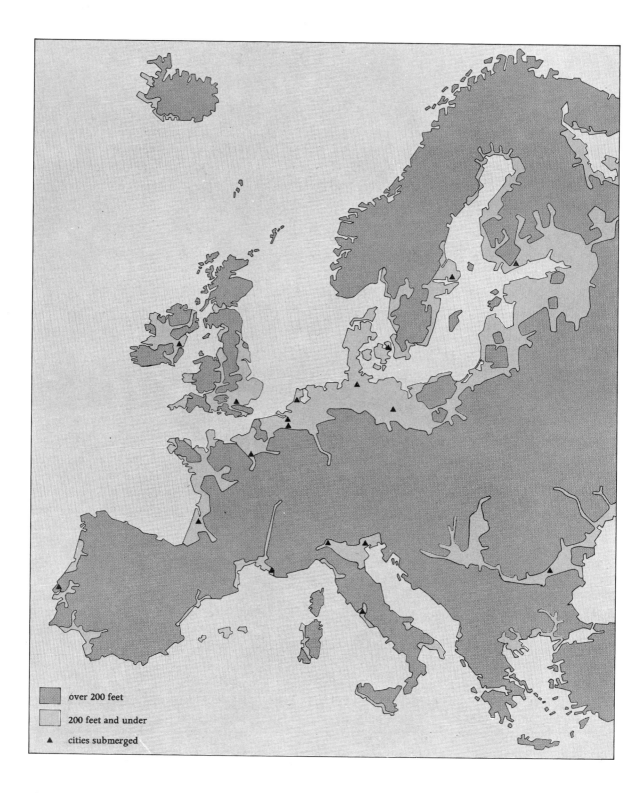

over 200 feet

200 feet and under

▲ cities submerged

about forests growing at the poles and tropical conditions at the latitude of present-day New York or London were based on a total misapprehension – the belief that the continents have stood rooted to their present spots since the world began. The confirmation that the continents have continually wandered about the globe completely alters the interpretation of coal seams found in Antarctica and tropical corals in Britain. When you make allowance for those movements, the climatic zones of the world turn out to be broadly similar to what they are today. But the formation of huge land masses that put Europe and eastern North America far from the sea at times and left them largely covered by desert complicates the interpretation.

When continents drift across the poles they tend to ice up. This happened on a large scale around 300 million years ago in Argentina, southern Africa, India and Australia, which at that time were all nestled together around the South Pole. The ice persisted for many millions of years. In an earlier episode, about 450 million years ago, northern Africa was near the South Pole. The evidence of ice sheets covering the deep Sahara at that time made no sense at all until continental drift was accepted as the explanation. And there were plenty of icy periods before then – at least a dozen altogether in the last 2600 million years, when various continents took turns in running into the long night of the polar winter.

All of this tells us that the stage was set for our present series of ice ages by the continents adopting their present-day positions. Long before the ice erupted in the northern lands Antarctica moved to the South Pole and began taking ice aboard. Antarctica was already close to its present position 50 million years ago, but at that time Antarctica and Australia were a single continent, New Zealand had recently broken from them, and the tip of South America was about to tear itself

from the tip of the Antarctic peninsula. These continental connections made a big difference – they prevented the establishment of the unimpeded vortices of wind and water that today swirl around the southern ocean and keep Antarctica in isolated deep-freeze. By 40 million years ago the rifts had occurred. With Antarctica now on its own, the whole world cooled rapidly: Nicholas Shackleton, examining the heavy oxygen in sea animals of that time, detects a sharp fall in temperature of the world's oceans about 40 million years ago.

New information about this early history of the present icy era comes from the remarkable American research ship *Glomar Challenger*. Her drilling rig reaches much deeper into the sea bed than do the piston corers used in CLIMAP and similar projects. *Glomar Challenger* spent the southern summer of December 1972 to April 1973 drilling 32 holes off Antarctica and to the south of Australia and New Zealand. She carried Australian, New Zealand, Canadian and Danish scientists as well as the Americans. When work was in progress close to the Ross Ice Shelf an attendant icebreaker was kept busy pushing icebergs away from *Glomar Challenger*. Incidentally she found natural gas (methane) in the continental shelf of the Ross Sea – another token of Antarctica's former existence in warmer climatic zones.

The onset of southern icing was pushed a long way back in time. *Glomar Challenger* recovered stony grains of the sort shed by melting icebergs, going back 70 million years, to the time of the demise of the dinosaurs. This finding lends a little sustenance to the theory that the dinosaurs were killed off by a world-wide change of climate – essentially a cooling – which may in turn have been due to the loss of carbon dioxide taken out of the air to make huge deposits of chalk.

The key conclusions from the long ocean-bed cores

were that the Circum-Antarctic current established itself 30 million years ago and by 20 million years ago Antarctica was heavily iced. The cores also told of a dramatic event, 5 million years ago, when the ice rested on the sea bed 250 miles beyond its present limits. The upwelling of water around Antarctica, just inside the Circum-Antarctic current, which made the southern ocean so rich in krill and whales, also began vigorously about 5 million years ago. By 4 million years ago the ice was melting back to something like its present extent, since when signs of great change are remarkable for their absence.

Meanwhile the North Atlantic was opening; Africa was driving into Europe to make the Alps, and India into Asia to make the Himalayas; the Gulf of Mexico was forming as North and South America linked together; the northern continents were encircling the Arctic Ocean. Only with a much better grasp of how land masses and mountains affect ocean currents and the global winds will the experts be able to say anything definite about the relative importance of these different northern features in causing the ensuing changes of climate. What they do know now is that icing began in Alaska 10 million years ago and there were glaciers in California and Iceland three million years ago. Soon after that – just when is still in doubt – the huge ice sheets of North America and Europe began their periodic performances.

Given the present motions of the continents, riding on the great plates of the Earth's broken shell, is there any sign of a break in the conditions that favour ice ages? The answer is no, not during the next 50 million years or so. On the contrary, with the Atlantic growing wider the benign influence of the Gulf Stream on Europe may weaken. Africa and Europe are driving together, destroying the Mediterranean and replacing it by a new chain of mountains of which Cyprus is the

first symptom. That could make northern Europe as chilly as Tibet, but more snowy. Antarctica shows no sign of abandoning the South Pole. Australia is edging towards the equator, where it will benefit from greater rainfall itself but will reduce the supply of warm, moist air to the weather machine as a whole. If our planet is ever to turn into that superfrozen ball of ice, an unlucky deployment of the continents will be the most likely reason for it.

The movements of continents have for the time being tuned the geography of the Earth to the point where the great oscillations of the ice sheets are possible. This explanation of the onset of the series of ice ages is so adequate that there is no room for another one that astronomers are trying to foist upon us. I would not even mention it, if it did not involve a fascinating mystery. For years an American physicist, Raymond Davis, has been trying unsuccessfully to detect ghostly particles called neutrinos coming from the Sun. They should be pouring out of the core of the Sun as the nuclear reactions proceed which are the source of its great heat. In a strange solar observatory a mile underground in the Homestake gold mine in South Dakota, Davis sets a trap for the neutrinos – a large tank filled with cleaning fluid. Thirty times a month, a neutrino ought to react with a chlorine atom in the cleaning fluid and alter it into an argon atom. In fact he finds no more than one such event a month.

Rather than scrap their theories, some astronomers want to say that the core of the Sun is stirring itself and cooling down for a few million years, like a fire being restoked. Series of ice ages would then be due to the peculiar stirring in the Sun every few hundred million years. While this story might help the astronomers out of a jam, the geologists don't need it. They would sooner hear about faster changes in the Sun that might explain short-term changes of climate on Earth.

Buried lands. The extent of the North American ice sheets 18,000 years ago has been mapped afresh for the CLIMAP project by Paul Mayewski of the University of Maine. It shows a corridor between the main Laurentide ice sheet and the ice on the western mountains.

A search for reasons

Until the recent discovery that ice ages are frequent and the warm periods in between have been brief, the search for the causes of ice ages was a leisurely pursuit. The geologists, ice scientists and climatologists discussed possible reasons for drastic changes of climate in the remote past and a supposedly remoter future as one might meditate upon the jawbone of a dinosaur. Dozens of theories and variations of theories vied with one another in a futile manner – futile because crucial evidence was either lacking or positively misleading. Today, a new sense of urgency prevails.

The next ice age may be due about now and either to forecast its onset or to devise ways of preventing it we ought to know the reason for it. A conclusion is still lacking, but not the determination to test the theories and try to settle the issue. The intricacy of the weather machine hampers the search for an explanation of the periodic freezing of the northern lands. Indeed the answer could be a conspiracy of several causes.

Suspended over this human enquiry into the ways of our planet is the possibility of a monstrous paradox. Heating, rather than cooling, could conceivably start an ice age. It could, for example, encourage Antarctic ice surges, of the kind discussed earlier, as a possible cause of ice ages or sudden coolings. A more general proposition is that the accumulation of unmelted snow, which makes the ice sheets, depends on the supply of snow as well as on its failure to melt. You need fairly warm, moist air that can manufacture snowflakes. The air should not be so warm as to melt the snow before it reaches the ground, but cold, dry Arctic air is a poor snow-maker. The idea is weak when applied in detail in the theory of a 'warm Arctic', which clears the northern seas of ice so that they can produce plenty of snow on the surrounding lands. North-eastern Siberia

and northern Alaska carry relatively little ice during an ice age; if the Arctic were a major source of snow they would have plenty. But the heating paradox is worth keeping in mind, if not as a mechanism for starting an ice age, at least as a local factor determining the situation of ice sheets downwind from warm ocean water.

Many weathermen would like to find the explanation for ice ages within the weather machine itself. The ordinary behaviour of air, water and ice may conceal the means of engineering an ice age, without needing any changes in external circumstances, such as the intensity of sunlight. The Antarctic ice surge and the 'warm Arctic' are two mechanisms on offer: the evidence is encouraging for neither. More generally, one can imagine the weather machine wandering by chance into a persistent block that maintains snow and cold on the northern lands. Once past a critical point, the effects of the incipient ice sheets on the weather patterns could reinforce the growth of the ice sheets. Then the weather machine has the problem of climbing out of the ice age again. If the oceans are thoroughly chilled, that may cut off the supply of snow to the ice sheets, while the supply of cold fresh water from melting ice floats on the ocean, freezes easily, and keeps the ocean cold while melting on land proceeds. One ingenious suggestion is that the wind-blown dust of the ice ages, the loess, dirties the ice sheets, helping them to absorb sunlight and so melt.

The possibility that the weather machine engenders its own ice ages entails a thought-provoking question: can I start an ice age by waving my arm? The air stirred by a casual gesture may grow into an eddy influencing the history of a thundercloud, which in turn modifies a depression, which helps to re-route the main jet stream, which produces a block and a very cold summer, which starts an ice age. Surprisingly, perhaps, nothing in the theory of the fluid turbulence of the atmosphere rules out such a sequence of events. On the contrary, the theory emphasises the possibility of very rapid growth of very small disturbances. Although much bigger disturbances like hurricanes and avalanches and jet aircraft are occurring all the time, their precise effects depend on the smaller disturbances among which they are operating. Until this issue is resolved, a prudent man might well keep his hands in his pockets.

For those who would explain the ice ages as internal quirks of the weather machine the chief difficulties come from the discovery of the 100,000-year ice-age cycle. You would expect less regularity, for one thing – more random timings and false starts. (Perhaps a sudden cooling is a false start.) The other difficulty is the time-scale, compared with the 'memory' of the various parts of the weather machine. The air wipes its slate clean about once a month; sea ice spreads and retreats from one season to the next; even the oceans are fully stirred every thousand years or so, erasing any unusual effects of warmth or cold. In principle you might have two independent cycles in the ocean, one producing maximum cold once in 1000 years and the other once in 990 years. Their cooling effects would coincide, producing maximum cooling, every 99,000 years. At present this is nothing more than an arithmetical curiosity. Only large ice sheets on land have an inherent persistence of many thousands of years – by which point one is begging the question of why they come and go, with the ice surge as a possible but unproven answer.

The same apparent lack of a natural rhythm to match the 100,000-year ice-age cycle is the chief drawback to the volcanic theory of ice ages. There is no particular difficulty about imagining a stupendous volcanic eruption, or series of eruptions, drawing a veil across the Sun and letting the ice start to form. But in that case the dust should be preserved in the ocean bed.

That the ice sheets themselves might modulate volcanic activity, releasing it when it melts and so developing a sort of rhythm, is not very plausible. Volcanoes are due to the plates of the Earth's shell pressing together or tugging apart, propelled by mighty forces originating deep inside the Earth and persisting for many millions of years. The weight of ice can, it is true, depress a continent into the Earth by as much as 3000 feet; the northern continents are still rising after being relieved of their most recent ice sheets. On the other hand, the lowering of the sea level during an ice age relieves the water pressure on the ocean bed. Most volcanoes, in any case, are in warm zones, whilst the great ice sheets lie mainly on volcanically quiet lands.

Variable sunshine

If you want to back a theory of ice ages that cannot be disproved in your lifetime, say that the Sun varies its output of energy in a cycle of 100,000 years. There is no obvious reason why it should not do so but it would be very hard to detect. Only above the atmosphere can you escape spurious variations in the apparent brightness of the Sun as the transparency of the air changes. So you would have to stage a long succession of satellite flights of identical instruments for measuring the Sun's brightness. After 50 years you might find variations corresponding to sunspot cycles and cycles of cycles (see Chapter 1) and that is a compelling reason for mounting just such a programme of satellite observations. But any change in the Sun that takes 100,000 years to complete would require many centuries of satellite observations to confirm or refute. The blinking-Sun theory looks fairly promising for the relatively small variations in climate that take a century or two but as an explanation for ice ages it is mere conjecture.

Very much more pertinent is the fact that, even if the

Sun burns steadily, the pattern of heating of our planet changes. The gravity of the Sun, the Moon and the other planets causes the Earth to wobble and vary its orbit around the Sun, over many thousands of years. These are the only known phenomena in nature with rhythms of the right sort of duration to compare with the rhythms of the ice ages. The sunshine falling on the northern lands in summer waxes and wanes. When it wanes the snow may fail to melt in summer, so causing an ice age. Alfred Wegener, the visionary of continental drift, was very interested in the theory, but the 'Milankovitch theory' is the usual name for it, from the Yugoslav geophysicist Milutin Milankovitch, who advocated it in the 1920s and 1930s. Until now everyone has been working with incorrect histories of the ice ages, so a great deal said about this theory during the past half century is outdated. With the new time-scale for the last eight ice ages climatologists are busily looking afresh at the variations in the Earth's planetary behaviour to see whether they fit the pattern of the advancing and retreating ice.

The ancient Egyptians planned the entrance passage of the Great Pyramid of Cheops so that the bright northern star Thuban would shine down it for ever more. Alas for their ever more, the star is now 20 degrees out of alignment. This is an instance of the changing astronomical circumstances which can affect the climate of the Earth profoundly.

The Earth's axis is not upright with respect to its orbit around the Sun. It inclines at an angle of 23·4 degrees and it tips the northern hemisphere towards the Sun in July and the southern hemisphere towards the Sun in January. The situation is not quite fair, because our planet's orbit around the Sun is not exactly circular. The Earth is three million miles closer to the Sun in January than it is in July, which means that the sunshine is weaker by 7 per cent during the northern summer than during the southern summer. But all of these factors can change.

Three different rhythms come into play, to the delight of cycle-mongers. First, the Earth's orbit can 'stretch' by departing much further from a circle, and then revert to almost true circularity. A complete cycle (near-circular to stretched and back to near-circular) takes 90,000 to 100,000 years, though both the duration and the extent of the stretch can vary. In extreme cases of stretching, the intensity of sunshine reaching the Earth may vary by up to 30 per cent from its strongest to its weakest, in the course of a year. That is a far greater change than anyone has suggested for variations in the power of the Sun itself. At present, though, the orbit is not extreme and is becoming nearly circular.

Much more rapid than the change in orbit is the change in the season of the closest approach to the Sun. Although it occurs during the southern hemisphere summer at present, 10,000 years ago the northern hemisphere was in summer when the Earth was closest to the Sun – as it will be 10,000 years hence. The reason is that the Earth wobbles like a spinning top and swivels its axis around and around. Astronomers call this behaviour the 'precession of the equinoxes' and it is responsible for the increasing misalignment of the Great Pyramid passage. The cycle-time is 21,000 years. The phase we are in at present is the worst for northern summer sunshine.

Finally, the Earth also rolls like a ship, altering the tilt of its axis between 21·8 degrees (more upright) and 24·4 degrees (more tilted). One complete roll, up and down again, takes 40,000 years. The greater the tilt, the more pronounced is the difference between winter and summer. For the last 10,000 years the Earth has been in the upward roll. That would mean (if this were the only factor) summers becoming less extreme – in other words, cooler.

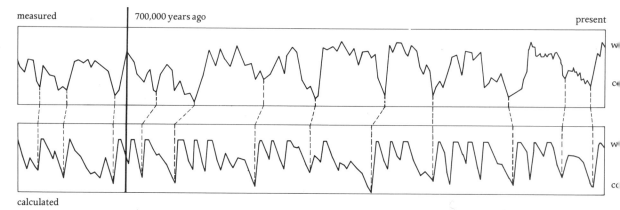

calculated

The cause of ice ages. The comings and goings of the ice are now reasonably well explained for the first time. The 'measured' record (N. Shackleton and N. Opdyke) shows the changing amount of heavy oxygen in marine fossils, which is proportional to the amount of ice in the world. A date marker at 700,000 years ago is provided by a reversal in the Earth's magnetism detected in the ocean-bed samples; the intervening dates are uncertain, because of variations in the rate at which mud accumulated on the ocean bottom. The second, 'calculated' graph has an exact time-scale. It is based on the variations in the energy of sunlight arriving in summer at latitude 50°N – sunshine needed to melt the winter snow. The assumption is that whenever the summer sunshine is two per cent stronger than at present, ice diminishes at a proportionate rate, up to a certain limit; when it is less than that, ice accumulates proportionately to the shortfall, at 0·22 of the rate of melting. The main increases and diminutions in the ice are accounted for. (Calculations by the author from A. D. Vernekar's tables.) The reasons for the variations in the summer sunshine are shown in the remaining diagrams. The antics of the Earth in its orbit, due to astronomical causes, affect its attitude and its distance from the Sun during the months of the northern hemisphere summer.

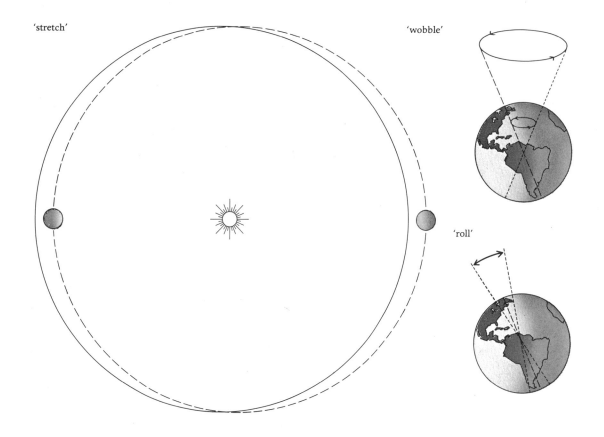

'stretch'

'wobble'

'roll'

To sum up, these are the Milankovitch cycles:

'stretch'	90,000-100,000 years
'wobble'	21,000 years
'roll'	40,000 years

Combining them, and correcting for variations due to complex astronomical causes, make for tricky arithmetic. When the sums are done, though, the past variations in sunshine in some climatic zones at some times of the year certainly bear more than a passing resemblance to the modern records of changing conditions, through the ice ages and the warmer periods in between, with the same sorts of peaks and troughs at the same intervals. The difficulties come with the details – here, perhaps, intense sunshine apparently coinciding with deep ice ages; there, low sunshine in periods reckoned to have been warm. Because the dates of the individual ice ages and warm periods during the past 700,000 years are still a matter of guesswork, you can slide them around a little to fit your sunshine patterns better. Some experts even favour using the sunshine calculations as a method of dating! And you can make different assumptions about the relative importance of 'stretch', 'wobble' and 'roll'.

You can declare that you want to explain, with the sunshine calculations, the onset of the ice ages, or the onset of the warm periods, rather than trying to match the intervening temperature records. And you can say, if you wish, that some other factor than the sub-Arctic summer sunshine is the really important thing. For example, reduced winter sunshine in the stormy zone of the northern hemisphere arguably promotes the accumulation of snow. The biggest changes in sunshine actually occur at the equator. There are more games to play when you introduce, as you should, some idea of the global wind patterns and their contradictory effects. In short, there is endless scope for fiddling – as demonstrated by the survival of the Milankovitch theory both before and after the overturn in man's knowledge of the frequency of ice ages, and the corrections to the time-scales.

Yet the Earth's orbit and orientation *have* changed rhythmically and the sunshine falling on different regions in different seasons *has* varied markedly. Theories of the Milankovitch kind are by far the most promising explanations of ice ages that we have. Exceptionally thorough calculations of the sunshine changes all over the world in winter and summer, for the past two million years, were published by Anandu Vernekar of the University of Maryland in 1972. Combined with the 700,000-year-old magnetic reversal, which has corrected the overall time-scale of the ice ages, they open the way for a new and perhaps definitive version of the Milankovitch theory.

Lacking any up-to-date comparison by specialists, of these astronomical and glacial data, I decided to attempt my own for the purposes of this book. From Vernekar's tables, it seems to me that the crucial zone is not at the Arctic Circle, as has often been supposed, but at 50°N (the latitude of Frankfurt and Seattle). This fits in well enough with the snowblitz theory and the discovery of the English Channel glacier. The diagram opposite shows a simple running calculation of the amount of ice in the world. It assumes that, if the summer sunshine at 50°N falls below a certain level (two per cent higher than at present) the volume of glaciers and ice sheets grows in proportion to the deficit. Summer sunshine above that level melts ice – four or five times as rapidly, up to the limit when the meltable ice has gone. The other part of the diagram shows the amount of ice in the world as measured by Nicholas Shackleton in his small sea shells.

Although the precise extent of the rise or fall at each step is not perfect by any means, the major variations are captured by the calculations. They also register known events not visible in the oxygen-isotope curve, namely the cooling at 90,000 years ago and a warming at 100,000 years ago. Dates deduced by Emiliani and Shackleton, which take account of variations in the rate at which sediment accumulates on the ocean floor, compare quite well with the dates I identify by matching particular cold periods visible in the two curves. My assumptions are almost frivolous and meteorological effects are notoriously complex, so the match between the wiggles and dates shown in the diagrams is very much better than it deserves to be – unless the Milankovitch theory is essentially correct.

On the face of it the Milankovitch cycles and internal behaviour of the weather machine offer ample resources for running the world either as a refrigerator or as a stove, in accordance with the newfound prehistoric record. So although the dust of dispute may not settle until the dates of all the alternations of warmth and cold are known with more precision, there is now good reason to hope for success in understanding ice ages. Meanwhile, there is at least a *prima facie* case for invoking the present and future effects of 'stretch', 'wobble' and 'roll' in any attempt to estimate when the next drastic change of climate may occur.

Reckoning the risks

According to the old ideas about ice ages, no really drastic change of climate could occur in our own lifetime. In 1970 an expert would have said:

'The ice sheets will take many thousands of years to form and spread. So even if the next ice age is now due – even if it has already begun – the worst that can happen in the next hundred years is that the world will become slightly colder – perhaps as bad as the Little Ice Age. The glaciers may advance a few hundred yards but otherwise we and our children won't know too much

about the coming ice sheets. With plenty of centuries to adjust to any change in that direction, human beings have more important things to worry about than the threat of ice.'

The reckoning of the risks changes completely when the sudden cooling and the snowblitz are taken into account. They tell us that the ice age could in principle start next summer, or at any rate during the next hundred years, with a ferocity that could not be mistaken for a 'mere' climatic fluctuation like the Little Ice Age.

In the 100,000 years of the typical ice-age cycle, even an onset in a hundred years is almost instantaneous. Going by past form, the warm periods between ice ages last about 10,000 years and ours has lasted 10,000 years. One might therefore argue that there is a virtual certainty of the next ice age starting some time in the next two thousand years. Then the odds are only about 20 to 1 against it beginning in the next 100 years. If we treat a sudden 1000-year cooling as an extra risk, over and above the ice age proper, the odds shorten to something like 10 to 1 against. If even roughly correct, that is a very high risk indeed for an event that could easily kill two thousand million people by starvation and delete a dozen countries from the map. Actuarially it means that the threat of ice reduces the life expectancy of everyone on the planet by several years.

That is crude arithmetic. What about the subtler arithmetic of the sunshine, taking account of the Milankovitch variations in the Earth's tilt and orbit? We are living in a less variable period, astronomically speaking, than the Earth has experienced during the last 1,600,000 years. In particular, the orbit has been close to circular for 50,000 years and will remain that way for the next 50,000 years. The overall variations in sunshine will therefore be comparatively small, and optimistic climatologists argue that there is no powerful reason for the next ice age to come for many

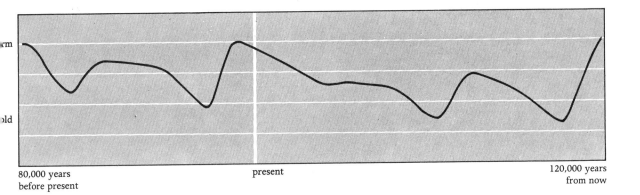

warm

cold

80,000 years
before present

present

120,000 years
from now

thousands of years. Unfortunately the changes in sunshine, though small, are not negligible.

During the past 10,000 years, since the Earth hauled itself out of the most recent ice age, the northern summer sunshine has diminished substantially, in the tropics and all the way to the pole. The last time it was so reduced was 20,000 years ago, when Canada and northern Europe were lying buried under the ice. And Willi Dansgaard points out that we are in just about the same condition, in respect of sunshine and a declining warm period, that the Earth was in 90,000 years ago when the sudden cooling came. In fact the summer sunshine is less than it was then, although not as weak as it was 70,000 years ago at the start of the ice age.

If I may revert to my own comparison (page 133) of the ice record of the past with simple arithmetic based on variations in summer sunshine at 50°N, the calculations can be extended into the future to arrive at an explicit forecast, albeit an approximate one. The resulting diagram above indicates ice amassing fairly slowly compared with some previous episodes, yet inexorably. By this picture, the next ice age has already begun. The snowblitz has still to hit us but, to fulfil the prediction of the diagram, huge areas would have to begin carrying a thin crust of ice before long.

The countries at risk from ice are shown in the next column. And if the climatologists are correct, who now think that cold conditions in the north are likely to be linked with a drier climate in the tropics, there is also a long list of countries in danger of worsening drought. Many of the same tropical areas may even now be partially drying out because of a resumption in the Little Ice Age in the north. The onset could still be gradual, giving plenty of time for human populations to adapt to, or combat, the changes, or it could be disastrously rapid. The evidence, though, for the episode of the sudden cooling and for the mechanism of the snow-

blitz favours a catastrophic view of the threat of ice.

Countries in danger of obliteration (complete or almost complete) by ice sheets

Canada, Greenland, Iceland, Irish Republic, United Kingdom, Norway, Sweden, Denmark, Finland, Switzerland, Liechtenstein, Nepal, Sikkim, Bhutan, New Zealand.

Countries in danger of extensive glaciation

USA, USSR*, Mexico, Colombia, Argentina, Chile, Netherlands, Germany (FDR and DDR), Poland, Austria, Afghanistan, China, Australia.
*USSR: danger of obliteration of Estonia, Latvia, Lithuania, Belorussia, Kirgizia, Tadzhikistan and large areas of RSFSR.

Countries in danger of severe drought* during the onset of an ice age

Mexico, Guatemala, Guyana, Surinam, French Guiana, Brazil, Argentina, Uruguay, Paraguay, Mauritania, Senegal, Guinea, Mali, Upper Volta, Ghana, Togo, Dahomey, Niger, Nigeria, Chad, Cameroun, Central African Republic, Congo, Zaire, Kenya, Tanzania, Angola, Namibia, Zambia, Rhodesia, South Africa, Botswana, Afghanistan, Pakistan, India, Bangladesh, China, Indo-China, Australia.
(*Following a listing by Rhodes Fairbridge)

Gods and frost-giants

We humans are creatures of the ice ages. Our forefathers evolved away from the apes as the ice built up in Antarctica and Alaska and as East Africa, the cradle, turned dry – dissolving the forests into open grassland

135

where a man might sprint. During the past two million years, the newly attuned geography of the Earth let loose a score of ice ages, although mankind lost count of them until 1969. The rapidly changing climate must have been abominable for *Homo erectus* and the first attempts at *Homo sapiens*, but it gave clear advantages to an adaptable, clever animal. The appearance of modern man in the Middle East 50,000 years ago co-incided with a temporary recession of the ice in the midst of the last ice age, doubtless causing some new climatic affront to life in that part of the world. And we began to wander, into the caves of chilly France for example, and across to America. As the ice withdrew from the northern lands, we and the forests followed in behind it.

The thought that the ice might come back lingered in the myths of the northerners. The benign gods of Norse mythology are perpetually at war with the giants, who represent the exigencies of the natural world. Freyr, the chief god of the Swedes, represents sunshine and fertility. By bribery, threats or curses, he has to win the Earth-maiden Gerth from the clutches of the frost-giants. In another story, when a giant builds a wall for the gods, the payment is to be the goddess of fertility, together with the Sun and the Moon; by the trick of sending a mare to distract the giant's stallion that carts the rocks, the gods succeed in delaying the work so that they can invoke a penalty clause and refuse payment.

The end of the world is to be heralded by a summer that is no summer. The bitter cold persists and the Sun gives neither light nor warmth, so that the winter is three winters long. 'The twilight of the gods' is a realistic scenario for the onset of the next ice age, in snow-blitz fashion. But a creed from a warmer zone drove out such pagan thoughts; white was not the fashionable colour of doom. Only in the 1970s of the Christian calendar did wise men again become mindful of the ice.

I have presented, in this book, many disparate facts and theories about short-term and long-term climatic changes. At the risk of confusing the reader I have tried to hint at the extraordinary complexities and paradoxes of the weather machine, some of which give grounds for optimism. But when all is said and done, the simplest argument of all is discouraging. It is that for 95 per cent of the past million years the world has been very much colder than it is today; on a shorter time-scale, the warm spell in the northern lands in 1920-50 was a quite exceptional 30 years compared with the chilly centuries of the Little Ice Age, which may not yet have ended. Any natural change of climate is far more likely to be for the worse than for the better.

For residual optimism we must look to the very human ability to rival nature, by generating heat and carbon dioxide and dust, which at present gives cause for concern because we do not know what we are doing to the climate. If nature does not act too quickly, and nothing very drastic happens during the next 100 years, we can reasonably expect to learn in that time even how to ward off an ice age. The technology, of which I gave some speculative examples in Chapter 2, is the lesser problem. Far greater is the prodigious intellectual and computing task of understanding the weather machine better than we do.

The most hopeful sign of the times is the vigorous international cooperation in the Global Atmospheric Research Programme (GARP). It was conceived as a programme aimed at better routine weather forecasting but then the emphasis grew on GARP's second objective, namely 'better understanding of the physical basis of climate'.

Among the regional investigations under GARP, the Polar Experiment (POLEX) is the one of most obvious climatic significance. The Russians invented it, as an Arctic project, in 1968. Soviet scientists have been busy in the Arctic for many years and the manned and un-manned meteorological stations that they have set on icebergs are justly famous. In 1973 a Russian party took over a new one for POLEX and called it *North Pole-22*. The basic objective of POLEX is to study the winter waxing and summer waning of the Arctic ice, and ways in which its formation and movements influence the weather and the sea and are in turn influenced by them. The region of the experiment extends into the North Pacific and North Atlantic, using ships. Meanwhile the Americans and Canadians were preparing what they called the Arctic Ice Dynamics Joint Experiment with similar objectives. The Americans also decided to join in POLEX, but called for more attention to climatic changes, past and present, in the Arctic region.

The biggest effort so far planned for GARP is FGGE (pronounced 'Figgy'), the First GARP Global Experiment. As a scientific project its scale is breathtaking, surpassing even the Atlantic Tropical Experiment described in the previous chapter. The idea is to observe the whole world's weather as thoroughly as possible for a year, probably starting in 1977, with intensified observations in midsummer and midwinter. The World Weather Watch system of weather stations will be the bedrock of the observational programme. That at once involves almost every nation on Earth. These weather stations will be supplemented by observations reported by commercial ships and aircraft and by various satellites, including no fewer than five 'geo-stationary' weather satellites provided by the USA, USSR, Japan and Europe. These will give round-the-world coverage. In addition, elaborate research programmes will fill in vital pieces of the story.

Big balloons are under development at the US National Center for Atmospheric Research for filling in the scanty observations of the tropical winds and weather. The balloons, 75 feet in diameter, will float 15 miles above the tropics, each carrying 64 instrument packages which they will release, one at a time, in response to a radioed command. Each dropsonde will contain, besides instruments for measuring pressure, temperature and humidity, a receiver for the very sensitive Omega radio navigation system: by this means, as it falls to the surface, the dropsonde will register its sideways movements in the tropical winds. The aim is to have more than a hundred of the elaborate 'carrier balloons' dispersed around the tropics during both of the intensive periods of FGGE. The POLEX stations in the Arctic will also be operating during FGGE. There are plans to set 150 automatic buoys adrift in the southern ocean around Antarctica, to observe a great region where satellites will be unable to measure surface conditions because of persistent cloudiness. As many research ships as can be mobilised will obtain simultaneous observations of water conditions across the world's oceans. And the world's most powerful computers available will be at work analysing all the observations.

Out of the primary meteorological work of FGGE will come information about the weather machine of great importance for climatic studies. There will be confirmation or modification of present ideas about how the global winds are generated and how the weather in one part of the world influences the weather in another. And perhaps for the first time there will come a reasonably good understanding of how the jet stream and surface conditions arrive at their compromises about what the weather is to be at each place determining whether a season is good or bad. 'Thermal forcing' is the phrase the scientists use for the persistent influence of lying snow, of sea ice and of warm and cold patches of ocean water. At present there are rules of thumb about their effects. Only with intensive and worldwide meteorological measurements, as in FGGE, is there any real hope

The Weather Machine

of understanding precisely how they determine the weather in places far away.

From 1972 onwards, such moves were afoot to strengthen the climatic studies in FGGE that some weathermen began complaining that people were forgetting about the first GARP objective – better weather forecasting. One move was to bring the computer model-makers together with experts on tree-rings and the like, at a conference in Sweden in the summer of 1974. There were also pressures, mainly from the United States, to add new observations to the FGGE programme, relevant to understanding climate. These included world-wide measurements of the incoming sunlight and its reflection from the land and clouds, snow and ice by land and sea, the water content of clouds and rivers, the effects of forest fires, far more elaborate observations of the near-surface waters of the oceans, and pollutants and other trace gases in the atmosphere.

I have deliberately ended with these mundane technicalities; a palaeolithic reader would have approved. In the 20th century we have built a fair-weather world. A boat may put to sea with every sail spread to try to catch the slightest puff of wind, while the crew sunbathes. Yet they know that before sunset they could be clawing off a lee shore under a rag of storm canvas, shivering in their oilskins and safety harnesses under drenching spray. They hope the storm may never happen but the boat is built and rigged and fitted out on the assumption that it will. Money and effort has been spent in the hope it will have proved unnecessary, just to minimise the consequences of a few hours' misfortune with the weather. The same cannot be said of our human civilisation. Our farms, economies, transport systems, even our very houses are scarcely proof against the cold, if that is what the trend is, and we have entered an era of expensive fuel.

If the trend also means persistent drought in the tropics, mankind may be in very deep trouble indeed, never mind the Little Ice Age or the Big Freeze. The supposedly crafty fellows of the ice age are in poor shape, demographically, politically, economically and scientifically, to react to any substantial change of climate, whether it be natural or man-made. Perhaps we shall be lucky. In the nick of time we may comprehend the problem and cope sensibly and humanely with unavoidable disasters like the African drought. We may even bury our political differences while we take action to arrest a dangerous change in the weather. But my dogged optimism flags when I think of one man's shipwreck being another man's harvest, and of climatologically well-informed nations gaining money or strategic advantage from other nations' disasters.

Further reading

Much of the content of this book is new and not described in previous books accessible to non-expert readers. But there are many aspects of meteorology and climatology that I have passed over lightly and the following books, among many others, would repay the reader who is anxious to pursue the subject further.

Modern Meteorology and Climatology by T. J. Chandler (Nelson; London 1972). A crisp, well-illustrated introduction.

Weather and Climate by R. C. Sutcliffe (Weidenfeld & Nicolson; London 1966). A book for the general reader about what meteorologists do.

Meteorological Challenges: A History edited by D. P. McIntyre (Information Canada; Ottawa 1972). A review by leading meteorologists on the centenary of the Canadian Meteorological Service.

The Changing Climate by H. H. Lamb (Methuen; London 1966). A collection of papers by an outstanding pioneer. Somewhat technical, but less so than Lamb's monumental *Climate: Present, Past and Future*, of which Volume I appeared in 1972 (Methuen; London).

Times of Feast, Times of Famine by E. Le Roy Ladurie (Allen & Unwin; London 1972. Doubleday; New York 1971). An historian's view of climatic change in Europe since the year 1000.

Inadvertent Climate Modification. Report of the Study of Man's Impact on Climate. (The MIT Press; Cambridge, Mass. and London 1971). An authoritative discussion of a much-debated issue.

The Monsoons by P. K. Das (Arnold; London 1972). This is mentioned as a good example of a short, semi-technical book on a limited subject.

Acknowledgments

Acknowledgment is due to the following for permission to reproduce illustrations:

Pages 2, 3 NOAA; 6 Antonello Proto: Oxfam; 8 NASA: Camera Press; 9 Camera Press; 10 Professor W. Dansgaard; 12 (left) Popperfoto, (right) Swiss National Tourist Office; 13 Alec Nisbett; opp. 16 Horst Munzig: Susan Griggs; opp. 17 The National Gallery, London; 18 (top) Mansell Collection, (bottom) The Library of Congress; 20 The National Maritime Museum; 23 J. Allen Cash; 25 NOAA; 27 J. Allan Cash; 29 Dr K. A. Browning; 30-1 Frank W. Lane; opp. 32 The National Maritime Museum: Michael Holford; opp. 33 Steve A. Tegtmeier/Randy Zipser; 36 Alec Nisbett; 37 US Forest Service; 39 Alfred Gregory: Camera Press; opp. 40 James H. Meyer; opp. 41 Michael Freeman; 43 NOAA; 44 Harry Miller: Camera Press; 47 Brûlé: Gamma/The John Hillelson Agency; 48 J. Allan Cash; 51 NOAA; 52 (top) NOAA/GATE, (bottom) Syndication International; 54 NOAA/GATE; 55 Mansell Collection; opp. 56 The Victoria & Albert Museum; opp. 57 NASA; 58 Novosti; 59 Alec Nisbett; 60 Ben Ross: Camera Press; 66 NOAA; 70 J. Allan Cash; opp. 72 Professor J. D. Woods; opp. 73 Black Star; 74 Radio Times Hulton Picture Library; 75 The Royal Greenwich Observatory; 79, opp. 80 Alec Nisbett; opp. 81 NCAR; 82-3 Camera Press; 84 (top) Al Miller: NOAA, (bottom) Novosti; 85 Frank W. Lane; 86 NOAA/NHRL; 87 NOAA/NESS; 89, 90 Popperfoto; 95 NOAA/GFDL; opp. 96 Dr T. van der Hammen; opp. 97 J. Allan Cash; 100 Alec Nisbett; 103 NOAA; 106 Professor C. Emiliani/Dr N. J. Shackleton/*Science*; 108, 109 Professor G. Kukla; 111 Dr T. van der Hammen; opp. 112 Michael Freeman; opp. 113 NASA; 114 Lamont-Doherty Geological Observatory; 115 CLIMAP; 119 Michael Freeman; 120 Michael Astor: Camera Press; 123 US Navy; 124-5 Camera Press.

Diagrams by Diagram.

Picture research by Angela Murphy.

Index